Mobile Oriented Future Internet (MOFI)

Mobile Oriented Future Internet (MOFI)

Architectural Designs and Experimentations

Special Issue Editor

Seok-Joo Koh

MDPI • Basel • Beijing • Wuhan • Barcelona • Belgrade • Manchester • Tokyo • Cluj • Tianjin

Special Issue Editor
Seok-Joo Koh
School of Computer Science and Engineering,
Kyungpook National University
Korea

Editorial Office
MDPI
St. Alban-Anlage 66
4052 Basel, Switzerland

This is a reprint of articles from the Special Issue published online in the open access journal *Electronics* (ISSN 2079-9292) (available at: https://www.mdpi.com/journal/electronics/special_issues/future_internet).

For citation purposes, cite each article independently as indicated on the article page online and as indicated below:

LastName, A.A.; LastName, B.B.; LastName, C.C. Article Title. *Journal Name* **Year**, *Article Number*, Page Range.

ISBN 978-3-03936-102-1 (Pbk)
ISBN 978-3-03936-103-8 (PDF)

Contents

About the Special Issue Editor

Seok-Joo Koh, he received BS and MS degrees in Management Science from KAIST in 1992 and 1994, respectively. He also completed his PhD in Industrial Engineering at KAIST in 1998. From August 1998 to February 2004, he worked at the Protocol Engineering Center in ETRI. Since March 2004, he has been with the School of Electrical Engineering and Computer Science in the Kyungpook National University as an Associate Professor. He has published over 25 international journal papers with IEEE, Elsevier, and Springer-Verlag. His current research interests include mobility control in future Internet, mobile SCTP, and mobile multicasting. He has also participated in the International Standardization, as an editor in the ITU-T SG13 and ISO/IEC JTC1/SC6.

Editorial

Mobile Oriented Future Internet (MOFI): Architectural Designs and Experimentations

Seok-Joo Koh

School of Computer Science and Engineering, Kyungpook National University, Daegu 41566, Korea; sjkoh@knu.ac.kr; Tel.: +82-53-950-7356

Received: 22 April 2020; Accepted: 22 April 2020; Published: 23 April 2020

1. Introduction

With the explosive growth of smart phones and Internet-of-Things (IoT) services, the effective support of seamless mobility for a variety of mobile devices and users is becoming one of the key challenging issues. It is expected that the future Internet will be evolved toward 'mobile-oriented' [1]. In the mobile-oriented future internet (MOFI) environment, there will be a large number of mobile devices and users with a variety of heterogeneous mobile, wireless, sensor and vehicular networks.

To provide mobility management, a lot of protocols have so far been proposed, which include host identity protocol (HIP) [2], identifier-locator network protocol (ILNP) [3] and quick UDP internet connection (QUIC) [4]. However, these works may not be enough to provide mobility management in the MOFI environment. We may consider many recently proposed technologies, such as software defined networking (SDN) [5]. It is noted that SDN can be effectively used to control the network traffics by separating the control function from the packet data delivery function in the network. We also need to consider a variety of mobile networks in the MOFI environment, which include flying ad-hoc networks (FANET) [6] and connected vehicular networks [7].

2. The Present Issue

This special issue consists of seven papers that discuss how to enhance mobility management and its associated performance in the MOFI environment. The first two papers deal with the architectural design and experimentation of mobility management schemes, in which new schemes are proposed and the real-world testbed experimentations are performed. The subsequent three papers focus on the use of SDN for effective service provisioning in the MOFI environment, together with the real-world practices and testbed experimentations. The remaining two papers discuss the network engineering issues in the newly emerging mobile networks, such as FANET and connected vehicular networks.

In Reference [8], to overcome the drawbacks of the existing centralized mobility management schemes, the MOFI architecture is designed, which includes the separation of data and control planes for getting an optimal data path, and the distributed identifier–locator mapping control for alleviating traffic overhead at a central agent. In this work, the validity of the MOFI architecture is evaluated through the experimentations over the European Union (EU)–Korea testbed network. Reference [9] proposes an enhanced mobility management scheme in the ILNP-based mobile networks, in which the identifier-locators (ID-LOCs) mappings are managed in the fully distributed manner by using a mobile dynamic domain name system (m-DDNS) server located in each domain.

The SDN technology can be used for performance enhancement in the MOFI environment. Reference [10] presents an SDN-based quality of experience (QoE) control scheme for dynamic adaptive streaming over HTTP/3 (DASH), using the quick UDP internet connection (QUIC) [4] for mobile devices and users. Reference [11] discusses the testbed experimentations of SDN using SmartX boxes that are distributed across multiple sites. Each SmartX box consists of several virtualized functions that are categorized into SDN and cloud functions. Multiple SmartX boxes are deployed and inter-connected through SDN in the distributed environments. Reference [12] discusses the resource management scheme in the mobile cloud environment, which exploits SDN to introduce a vendor-agnostic concept.

The MOFI environment may consist of various types of emerging mobile networks, such as FANET and connected vehicular networks. FANET is an ad-hoc network for data transfer among the unmanned aircrafts in the three-dimensional space. Reference [13] proposes a routing scheme for FANET which can adapt to rapid changes in network topology and effectively improve the network performance. In Reference [14], the connected vehicular networks are considered, in which a new intersection signal control model is proposed based on vehicle-to-infrastructure (V2I) communication, and the simulation analysis is made with the real-world data.

3. Future

A variety of research works have so far been made with some testbed experimentations in the MOFI environment, as addressed in this special issue. However, some challenges still remain for further study in the future. New architectural models for MOFI need to be investigated by considering a variety of mobile devices/users for IoT services. The existing mobility management protocols, such as HIP and ILNP, can be reviewed and compared to design a more effective mobility management scheme. Some more testbed experimentations are also required for validation of the schemes for MOFI in real-world networks. The relevant standardization activities need to be investigated and initiated, if necessary, in the associated standards-defining organizations, such as ITU-T, ISO, JTC1, IETF, etc.

Author Contributions: S.-J.K. managed the whole editorial process of the special issue, 'Mobile Oriented Future Internet (MOFI): Architectural Designs and Experimentations', published by journal Electronics. He also drafted this editorial summary. The author has read and agreed to the published version of the manuscript.

Acknowledgments: First of all, I would like to thank all researchers who submitted articles to this special issue for their excellent contributions. We are also grateful to all reviewers who helped in the evaluation of the manuscripts and made very valuable suggestions to improve the quality of contributions. We would like to acknowledge the editorial board of Electronics, who invited me as a guest editor to this special issue. We are also grateful to the Electronics Editorial Office staff who worked thoroughly to maintain the rigorous peer-review schedule and timely publication.

Conflicts of Interest: The author declares no conflicts of interest.

References

1. Kim, J.-I.; Jung, H.; Koh, S.-J. Mobile Oriented Future Internet (MOFI): Architectural Design and Implementations. *Etri. J.* **2013**, *35*, 666–676. [CrossRef]
2. Moskowitz, R.; Nikander, P.; Jokela, P.; Henderson, T. Host Identity Protocol. *IETF Request Comments (RFC)* **2008**, *5201*, 1–102.
3. Atkinson, R.J.; Bhatti, S.N. Identifier-Locator Network Protocol (ILNP) Architectural Description. *IETF Request Comments (RFC)* **2012**, *6740*, 1–53.
4. Iyengar, J.; Thomson, M. QUIC: A UDP-Based Multiplexed and Secure Transport, IETF Internet Draft, Draft-IETF-Quic-Transport-27. 2020. Available online: https://www.ietf.org/id/draft-ietf-quic-transport-27.txt (accessed on 30 March 2020).
5. Xia, W.; Wen, Y.; Foh, C.H.; Niyato, D.; Xie, H. A Survey on Software-Defined Networking. *IEEE Commun. Surv. Tutor.* **2015**, *17*, 27–51. [CrossRef]
6. Bekmezci, I.; Sahingoz, O.K.; Temel, S. Flying Ad-Hoc Networks (FANETs): A survey. *Ad Hoc Netw.* **2013**, *11*, 1254–1270. [CrossRef]
7. Lu, N.; Cheng, N.; Zhang, N.; Shen, X.; Mark, J.W. Connected Vehicles: Solutions and Challenges. *IEEE Internet Things J.* **2014**, *1*, 289–299. [CrossRef]
8. Kim, J.-I.; Choi, N.-J.; You, T.-W.; Jung, H.; Kwon, Y.-W.; Koh, S.-J. Mobile-Oriented Future Internet: Implementation and Experimentations over EU–Korea Testbed. *Electronics* **2019**, *8*, 338. [CrossRef]
9. Gohar, M.; Choi, J.-G.; Ahmed, W.; Rahman, A.U.; Muzammal, M.; Koh, S.-J. Distributed Identifier-Locator Mapping Management in Mobile ILNP Networks. *Electronics* **2020**, *9*, 58. [CrossRef]

10. Guillen, L.; Izumi, S.; Abe, T.; Suganuma, T. SAND/3: SDN-Assisted Novel QoE Control Method for Dynamic Adaptive Streaming over HTTP/3. *Electronics* **2019**, *8*, 864. [CrossRef]
11. Risdianto, A.C.; Usman, M.; Kim, J.-W. SmartX Box: Virtualized Hyper-Converged Resources for Building an Affordable Playground. *Electronics* **2019**, *8*, 1242. [CrossRef]
12. Abbasi, A.A.; Al-qaness, M.A.A.; Elaziz, M.A.; Hawbani, A.; Ewees, A.A.; Javed, S.; Kim, S. Phantom: Towards Vendor-Agnostic Resource Consolidation in Cloud Environments. *Electronics* **2019**, *8*, 1183. [CrossRef]
13. Hong, J.; Zhang, D. TARCS: A Topology Change Aware-Based Routing Protocol Choosing Scheme of FANETs. *Electronics* **2019**, *8*, 274. [CrossRef]
14. Ren, C.; Wang, J.; Qin, L.; Li, S.; Cheng, Y. A Novel Left-Turn Signal Control Method for Improving Intersection Capacity in a Connected Vehicle Environment. *Electronics* **2019**, *8*, 1058. [CrossRef]

 *electronics*

Article

Mobile-Oriented Future Internet: Implementation and Experimentations over EU–Korea Testbed

Ji-In Kim [1], Nak-Jung Choi [2], Tae-Wan You [3], Heeyoung Jung [3], Young-Woo Kwon [2,*] and Seok-Joo Koh [3]

[1] Research Institute, Sillasystem Co. Ltd., Daegu 41566, Korea; jiin16@gmail.com
[2] School of Computer Science and Engineering, Kyungpook National University, Daegu 34129, Korea; peaceful7007@gmail.com
[3] Electronics and Telecommunication Research Institute, Daejeon 41566, Korea; twyou@etri.re.kr (T.-W.Y.); hyjung@etri.re.kr (H.J.); sjkoh@knu.ac.kr (S.-J.K.)
* Correspondence: ywkwon@knu.ac.kr

Received: 12 November 2018; Accepted: 12 March 2019; Published: 20 March 2019

Abstract: Today's mobility management (MM) architectures, such as Mobile Internet Protocol (IP) and Proxy Mobile IP, feature integration of data and control planes, as well as centralized mobility control. In the existing architecture, however, the tight integration of the data and control planes can induce a non-optimal routing path, because data packets are delivered via a central mobility agent, such as Home Agent and Local Mobility Anchor. Furthermore, the centralized mobility control mechanism tends to increase traffic overhead due to the processing of both data and control packets at a central agent. To address these problems, a new Internet architecture for the future mobile network was proposed, named Mobile-Oriented Future Internet (MOFI). The MOFI architecture was mainly designed as follows: (1) separation of data and control planes for getting an optimal data path; (2) distributed identifier–locator mapping control for alleviating traffic overhead at a central agent. In this article, we investigate the validity of the MOFI architecture through implementation and experimentations over the European Union (EU)–Korea testbed network. For this purpose, the MOFI architecture is implemented using OpenFlow and Click Modular Router over a Linux platform, and then it is evaluated over the locally and internationally configured EU–Korea testbed network. In particular, we operate two realistic communication scenarios over the EU–Korea testbed network. From the experimentation results, we can see that the proposed MOFI architecture can not only provide the mobility management efficiently, but also support the backward compatibility for the current IP version 6 (IPv6) applications and an Internet Protocol network.

Keywords: mobility management; architecture; implementation; experimentation; EU–Korea testbed

---

1. Introduction

As the current Internet architecture was designed for fixed network environments regardless of mobile network environments, future Internet architectures for the emerging network environments are widely discussed in recent research. Of those discussed, the incremental and clean-state approaches mainly dominate the future Internet research. In the incremental approach, one state is moved to another state with incremental patches, while, in the clean-slate approach, all the network stacks are redesigned from scratch to offer better abstractions and improved performance, as well as providing similar functionality based on new core principles [1]. In the past, Internet was wildly successful using the incremental approach. However, due to the rapidly emerging mobility technologies, today's Internet architecture faces many challenges. As a result, the clean-slate approach began receiving much attention to design the future Internet for mobile environments. However, applying the clean-slate approach to the current Internet infrastructure still incurs a deployment burden that requires the replacement or update of all network devices including routers, switches, and even hosts. As a result, in South Korea, research activities on future Internet architectures focus on Mobile-Oriented Future Internet (MOFI) [2], which is a new mobility management architecture based on the incremental approach. The great advantage of employing the incremental approach is that the new architecture and Internet services developed on the new architecture can be easily deployed over the current Internet infrastructure.

The MOFI architecture has three architectural components as follows: (1) host identifier and local locator (HILL), (2) query-first data delivery (QFDD), and (3) dynamic and distributed mapping system (DMS). Specifically, in HILL, each host has a globally unique host identifier (HID) for end-to-end communications, whereas the locator (LOC) of a network router is locally used for packet delivery. In QFDD, a location query is first executed before data delivery to obtain an optimal path between two connected hosts. In DMS, the mapping information between hosts is managed in a dynamic, distributed way. In order to provide compatibility with the existing Internet infrastructure, a host's IP address becomes a host identifier, and an access router's IP address is used as a locator.

Because the MOFI architecture only modifies the network devices used as a switch and a regional gateway for the data plane and a controller for the control plane, this design choice has a great advantage for deployment. Specifically, the proposed MOFI architecture can operate in the existing Internet Protocol version 6 (IPv6) Internet environment without any modification of the existing network infrastructure through LOC-based communications. Moreover, HID-based application services also can be used as is by utilizing the existing network infrastructure. As a result, while other architectures based on the clean-slate approach require the development and deployment burden of necessary devices and application services, the newly proposed MOFI architecture based on the incremental approach does not require any development of necessary application services and devices, thereby enabling fast deployment of Internet services in the new Internet infrastructure.

To evaluate the superiority of the newly proposed architecture, the architecture needs to be assessed through a set of simulations using NS3 [3] or OPNET [4] or real experiments on testbeds. Considering the scale of the Internet, the architecture needs to be evaluated on large-scale testbeds rather than simulations. Furthermore, because the MOFI architecture was developed in an incremental way, it must ensure that the new architecture can provide compatibility between existing Internet protocol stacks. To that end, we implemented the MOFI architecture on top of a Linux platform and then constructed a testbed across Korea and the European Union (EU) for the evaluation. More specifically, the data plane of the MOFI architecture was implemented using OpenFlow [5] and the Click Modular Router [6]. The control plane of the MOFI architecture was implemented using the OpenFlow, Click Modular Router, and UDP (User Datagram Protocol). This global testbed was established for the verification of the MOFI architecture.

The rest of this article is organized as follows: Section 2 presents the technical background that motivates our research. Section 3 summarizes an overview of the MOFI architecture. Section 4 presents the implementation details of the MOFI architecture and the globally constructed testbed between EU and Korea. Section 5 describes service scenarios and discusses the result demonstrated on the testbed. Section 6 concludes this article.

2. Background

In the last decade, both incremental and clean-state approaches dominated the future Internet research. In the incremental approach, a future Internet architecture is developed step by step based on the prior Internet architecture and infrastructure and, thus, existing Internet infrastructures and services can be used without any modification. The Internet was wildly successful using the incremental approach, as shown in the example of Mobile IP [7,8].

On the other hand, in the clean-state approach, an Internet architecture is newly designed and developed so as to maximize performance benefits. For example, the following research activities, including 4WARD [9–11], FIND (Future Internet Design) [12], MobilityFirst [13–15], GENI 9 (Global Environment for Network Innovations) [16], NDN (Named Data Networking) [17] were conducted based-on the clean-slate approach. 4WARD is an EU-initiated project that employed the concept of network virtualization, and a total of 37 partners were involved. FIND and GENI are NSF (National Science Foundation)-initiated projects to develop a new future Internet architecture. Through the GENI project, a new infrastructure was provided, and, through the FIND project, the proposed architectures were implemented and tested. To support new Internet features such as multicast, anycast, multi-path, and context-aware services, the MobilityFirst architecture employed a clean-slate approach. More recently, the NDN project was proposed to overcome the weakness of the current Internet architecture and to provide emerging communication patterns.

However, applying the clean-slate approach to the current Internet infrastructure requires additional development and deployment efforts. Thus, when moving toward future Internet, it is challenging to determine the transitioning time that meets all the requirements of a newly designed Internet architecture. In this article, we report our effort to construct a realistic testbed across the EU and South Korea. In addition, we tested the MOFI architecture implemented in an incremental way. In the discussion below, we describe our MOFI implementation and testbed construction in detail.

3. MOFI Architecture: Overview

3.1. Architectural Features

The Mobile-Oriented Future Internet (MOFI) architecture is an enhanced mobility management architecture that solves the problems that the current Internet faces. Table 1 shows the comparison of the current Internet's problems and MOFI design principles.

Table 1. Internet problems versus Mobile-Oriented Future Internet (MOFI) design principles.

Problems of Current Internet	MOFI	
	Design Principles	Functional Blocks
Internet Protocol (IP) address as both identifier (ID) and locator (LOC)	Separation of host ID (HID) and locator (LOC)	Host identifier (HID) and local locator (LOC) (HILL)
Address-based communication and global routing	HID-based communication and LOC-based local routing	
Data-driven packet delivery with non-optimal routes	LOC query before data delivery for optimal routes	Query-first data delivery (QFDD)
Static and centralized ID–LOC mapping system	Distributed HID–LOC mapping management	Distributed mapping system (DMS)

In the identifier–locator structure, the MOFI architecture uses the IP address of a host as host ID (HID), the media access control (MAC) address of the switch (SW), and the IP address of the regional gateway (R-GW), in which the host is attached as a locator (LOC).

Figure 1 shows a protocol model for the data delivery in MOFI. In this figure, the network layer of MOFI is divided into the communication and delivery layer. The communication layer can be implemented as a shim layer protocol between the transport and network layer. The HID field used for end-to-end communication between two end hosts is contained in the identity header. The delivery layer is divided into access delivery protocol (ADP) and backbone delivery protocol (BDP), which are used to deliver data packets between end hosts. For intra-domain data delivery, each SW translates ADPs. During this process, the identity header containing source and destination HIDs can be referred to by the SW, in which the LOC query operation of DMS is executed. For the inter-domain data delivery across different domains, each R-GW translates ADP to BDP. During this process, the identity header is referred to by R-GW, in which the LOC query operation of DMS is performed.

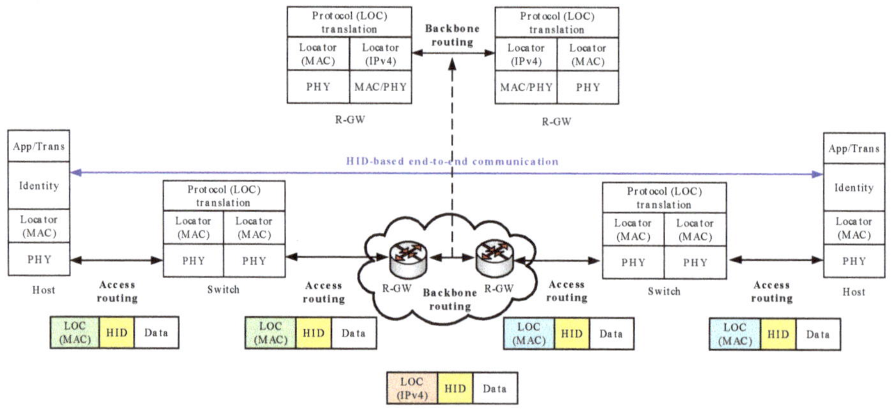

Figure 1. Protocol model for data delivery.

The data-driven packet delivery model used by current mobility protocols can induce non-optimal routes. In MOFI, therefore, we adopted the query-first data delivery approach, in which the LOC query operation is performed before transmitting data to find an optimal route between the two communicating hosts. Figure 2 compares the data-driven packet delivery and the query-first data delivery used in our approach.

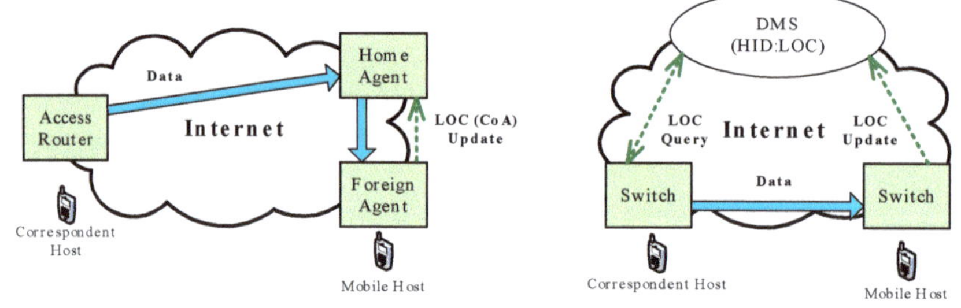

(a) Data-driven Packet Delivery with Non-optimal Route   (b) Query-First Data Delivery with Optimal Route

Figure 2. Data-driven packet delivery versus query-first data delivery.

In the data-driven packet delivery depicted in Figure 2a a mobile host (MH) updates its care of address (CoA) with LOC at the home agent (HA), when attached to a foreign agent (FA). The correspondent host (CH) sends data packets to the HA, which forwards these packets to the MH via the FA. However, this delivery mechanism can induce a non-optimal route. In Figure 2b, the HID and LOC of the MH are registered with DMS in the mobile environment. When the CH sends a data packet to the MH, a switch on the CH's side finds the LOC of the MH using the LOC query operation with DMS. Then, finally, the data packet is directly delivered to the MH. This data delivery mechanism provides better routing paths.

Figure 3 shows a hash-based distributed HID–LOC mapping management model used in the MOFI architecture. In the figure, each domain has its distributed mobility controller (DMC) for the mapping management information of mobile hosts. A selected DMC (S-DMC) is determined for each host using a hash function; for example, a simple modulo operation (%) can be used to determine the S-DMC for a host, such as "*(HID of the host) % (the number of DMCs in the Internet)*". Once S-DMC is determined for a specific host, the associated HID–LOC mapping information for the host will be maintained by the S-DMC.

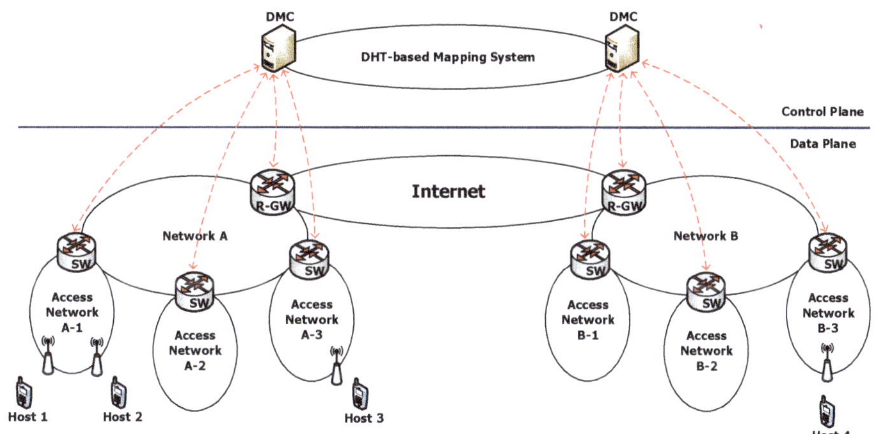

Figure 3. Distributed host identifier–locator (HID–LOC) mapping management in Mobile-Oriented Future Internet (MOFI) architecture.

Table 2 gives an overview of caches and registers used in MOFI. For each data and control plane, MOFI uses the following caches and registers: local binding cache (LBC), data forwarding cache (DFC), local mapping register (LMR), and serving mapping register (SMR). In the control plane, DMC maintains an HID–LOC mapping table (i.e., LMR) for its local host and SMR containing the domain information associated with each HID. In the data plane, SW and R-GW maintain the DFC that is updated by an LOC query operation for data forwarding. To operate HID–LOC mapping control, the operation is classified into the two operations: HID–LOC binding operations and LOC query operation for data delivery. These operations are described in the upcoming sections.

Table 2. Caches and registers. SW—switch; R-GW—regional gateway; DMC—distributed mobility controller.

Category	Entity	Cache/Register	Usage
Data plane	SW	Local binding cache (LBC)	Data forwarding (Host $\leftrightarrow$ SW)
		Data forwarding cache (DFC)	Data forwarding (SW $\leftrightarrow$ SW, SW $\leftrightarrow$ R-GW)
	R-GW	Data forwarding cache (DFC)	Data forwarding (R-GW $\leftrightarrow$ R-GW)
Control plane	DMC	Local mapping register (LMR)	HID–LOC mapping control (intra)
		Serving mapping register (SMR)	HID–LOC mapping control (hash-based)

3.2. HID–LOC Mapping Control Operations

In MOFI, the HID–LOC mapping control is divided into two operations including HID–LOC binding and LOC query. Furthermore, the LOC query operates in two modes—intra- and inter-domain. Thus, in this article, we discuss the three following cases: an HID–LOC binding operation, an LOC query operation in intra-domain, and an LOC query operation in inter-domain.

With the network attachment of a host, the HID–LOC binding operation is initiated. As described in Figure 4, during the binding operation, the HID of the host is registered with the SW attached to the host. Then, the SW updates the LBC with a message received from the host to record the HID of the host. Then, the SW sends an HID binding request (HBR) message to the DMC that the host belongs to. At the same time, the DMC receives the HBR message from the SW. The DMC updates its LMR and finds the selected DMC (S-DMC) that is selected for the host. In the case that the DMC becomes the S-DMC, the DMC sends an HBR message to the S-DMC. This HBR message contains the HID of the host and the LOC of SW. After successful HID–LOC binding, the S-DMC updates its SMR and responds to the DMC with an HID binding acknowledgement (ACK) (HBA) message, which is also forwarded to the host through the SW.

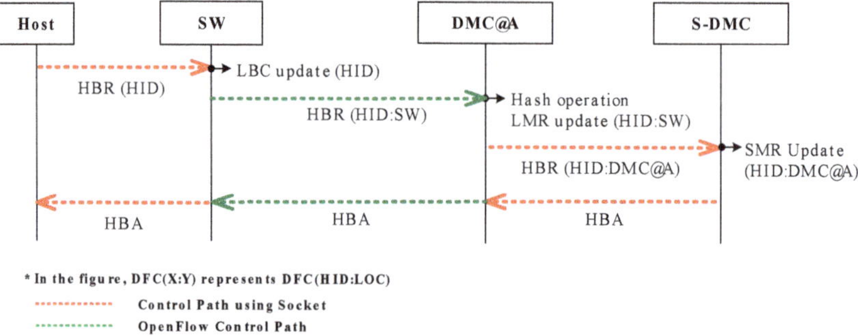

Figure 4. HID–LOC binding operations.

Figure 5 shows the intra-domain LOC query operations for data delivery. Once a data packet arrives from send host (SH), the SW (SW@A-1) sends an LOC query request (LQR) message to the DMC. Then, the DMC finds the receive host (RH)'s S-DMC using a hash function, and forwards the LQR message to the S-DMC. Upon receiving the LQR message from the DMC, the S-DMC looks up the SMR and responds to the DMC using the LOC query ACK (LQA) message. When the DMC receives the LQA message from the S-DMC, the DMC forwards it to the SW (SW@A-3) which belongs to the RH. When the SW of the RH receives the LQR message from the SW of the SH, the SW of the RH updates its DFC and looks up the LBC. After the LBC look-up, the SW of the RH responds to the SW of the SH through the DMC. When receiving the LQA message, the SW of the SH updates its DFC. Finally, the SW of the SH can exchange data packets with the SW of the RH through an optimal path.

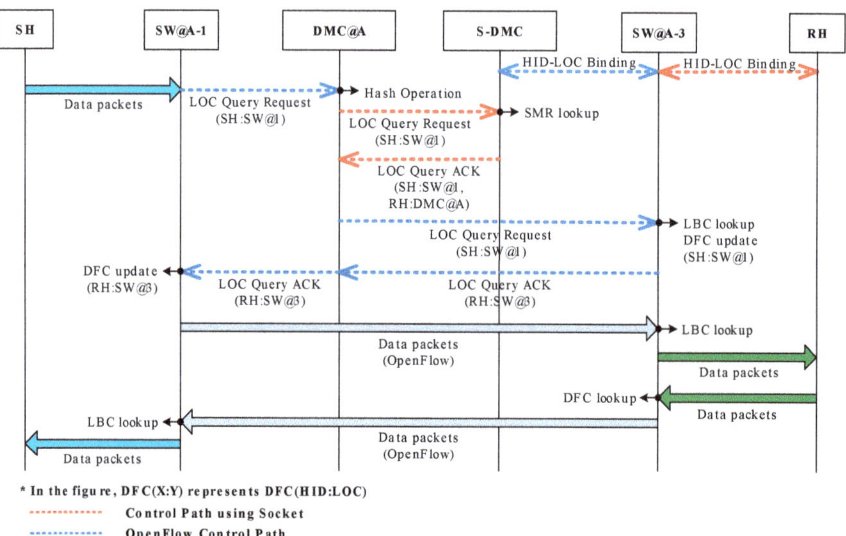

Figure 5. Intra-domain LOC query operations for data delivery.

Figure 6 shows inter-domain LOC query operations for data delivery, in which the RH exists in its own network domain with the S-DMC. The inter-domain LOC query operation is the same as the intra-domain LOC query operation until the SH's DMC sends an LQR message to the S-DMC. On receiving the LQR message, the S-DMC looks up the LMR and recognizes the existence of the RH in the same network. In this case, the S-DMC is the same as the DMC of the RH and sends an LQR message to the RH's SW. Receiving the LQR message from the RH's DMC, RH's SW looks up the LBC and updates the DFC with the received LQR message. The RH's SW responds to the LQA message to the RH's DMC. When the RH's DMC receives the LQA message, the RH's DMC sends an LOC update request (LUR) message to its R-GW to update the R-GW's DFC. Then, the R-GW updates its DFC and responds to its DMC by sending an LOC update ACK (LUA) message. As a result, both data and control planes can be completely separated. Then, the RH's DMC sends an LQA message to the SH's DMC. Once the SH's DMC receives the LQA message, the SH's DMC can exchange the LUR and LQA messages with its GW to update the DFC. After that, the SH's DMC sends the LQA message to the SH's SW. Upon receiving the LQA message, the SH's SW updates its DFC. Finally, the SH's SW can exchange data packets being sent to the RH's SW through the optimal path that includes the R-GWs.

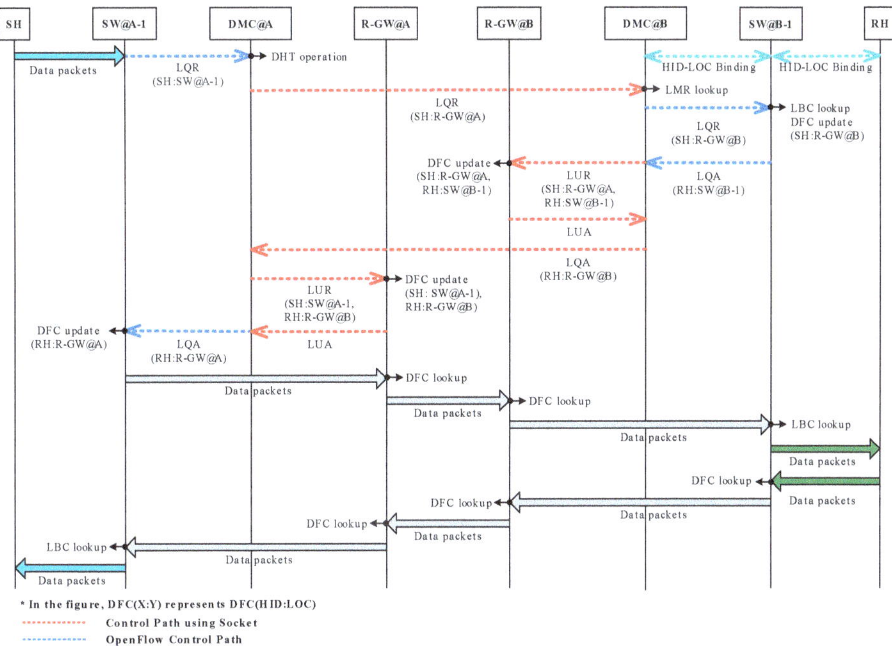

Figure 6. Inter-domain LOC query operations for data delivery (case 1).

Figure 7 shows inter-domain LOC query operations for data delivery. In this case, the RH and S-DMC exist in different network domains. The inter-domain LOC query operation is the same as the intra-domain LOC query operation until the SH's DMC receives an LQA message from the S-DMC. Upon receiving the LQA message, the SH's DMC sends an LQR message to the RH's DMC. When the RH's DMC receives the LQR message from the SH's DMC, it is the same as the inter-domain LOC query operations, in which the RH exists in its own network with the S-DMC.

3.3. Data and Control Packets

In MOFI, the HID is constructed with 2 bytes of a prefix, 4 bytes of a domain identifier, and 10 bytes of a subscriber identifier, as shown in Figure 8. The prefix field is not used in the current implementation. The domain identifier field is used for identifying a domain associated with the HID or a host. MOFI uses an autonomous system number (ASN) as a domain ID. For a 4-byte representation of a legacy 2-byte ASN, the first 2 bytes are set to "0" [18]. A subscriber identifier is allocated to each host by a domain administrator. Each domain can use this field as a "sub-domain ID" depending on its policy for HID management.

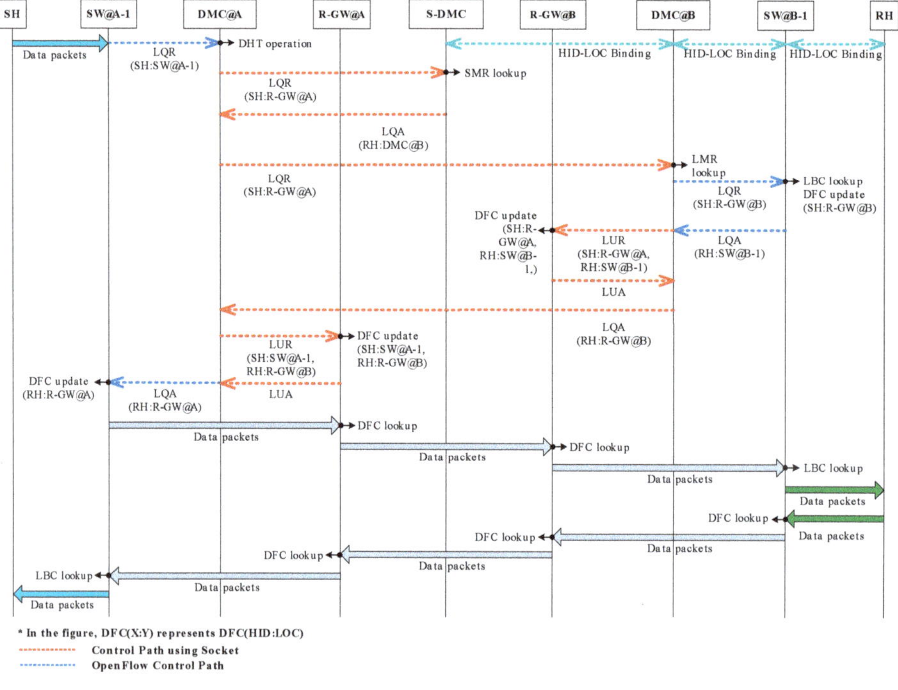

Figure 7. Inter-domain LOC query operations for data delivery (case 2).

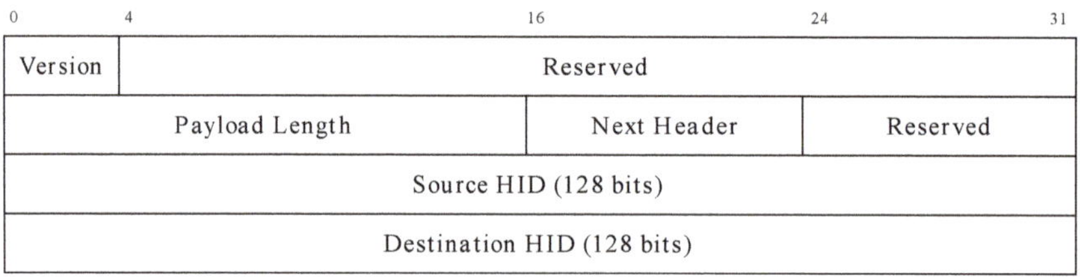

Figure 8. HID format.

The identity header is newly defined for the end-to-end data communication between two hosts. It is not responsible for data delivery or routing. That is, the HID contained in the identity header is not used for routing data packets in the network. Instead, it is used for the end-to-end communication using an upper layer transport layer protocol and a socket interface with an application program. In MOFI, the identity header format was designed to provide backward compatibility with the current IPv6 header, as shown in Figure 9.

Version	Reserved		
Payload Length		Next Header	Reserved
Source HID (128 bits)			
Destination HID (128 bits)			

0   4   16   24   31

Figure 9. Identity header format.

The identity header is similar to the current IPv6 header. The only difference is the absence of a traffic class, flow label, and hop limit. Instead, those fields are reserved. The version field is compatible with the current IP version. Next, the reserved field is set to 0. The payload length field is the length of the payload (in bytes) following this identity header. The next header field is the same with the next header of thIPv6 header. Furthermore, the second reserved field is reserved for the future use. The source HID and destination HID fields are used for HIDs of the source and destination.

Figure 10 shows the structure of data packets. For data delivery, each host constructs a data packet with the identity. For intra-domain access network delivery, an original packet is encapsulated by adding the ADP header. For inter-domain backbone network delivery, each R-GW translates an ADP header into a BDP header. In this article, the format of the ADP header is a MAC header, while the format of the BDP header is an IPv4 header.

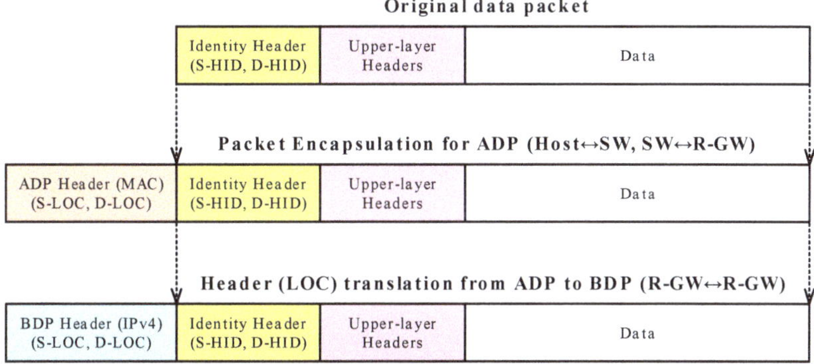

Figure 10. Structure of data packets.

In the control plane, there are six packets for HID–LOC mapping control. Table 3 shows the list of the control messages used in MOFI.

Table 3. Control messages. ACK—acknowledgement.

Message	Full Name	Encoding	From	To
HBR	HID Binding Request	0000 0000	Host or R-GW	R-GW or DMC
HBA	HID Binding ACK	0000 0001	DMC or R-GW	R-GW or Host
LQR	LOC Query Request	0000 0010	SW or DMC	DMC or SW
LQA	LOC Query ACK	0000 0011	SW or DMC	DMC or SW
LUR	LOC Update Request	0000 0100	DMC	R-GW
LUA	LOC Update ACK	0000 0101	R-GW	DMC

In MOFI, HBR and HBA messages are used for updating HID binding information between a host and R-GW. The LQR and LQA messages are used for LOC query operations for finding an optimal route. The LUR and LUA messages are used for updating the DFC for data delivery.

Each control message is encapsulated into a transport layer protocol that has a 20-byte common header and variable length parameters, as illustrated in Figure 11.

Common Header (20-byte)	Parameters (Variable Length)

Figure 11. Structure of the control message.

Figure 12 shows the common header format. The message type is an encoding value of the message shown in Table 3. The flag is used for various flags, which are described for each message.

The total length field is the length of the message in bytes, including a common header and parameters. The HID field is the HID of the host associated with the corresponding message.

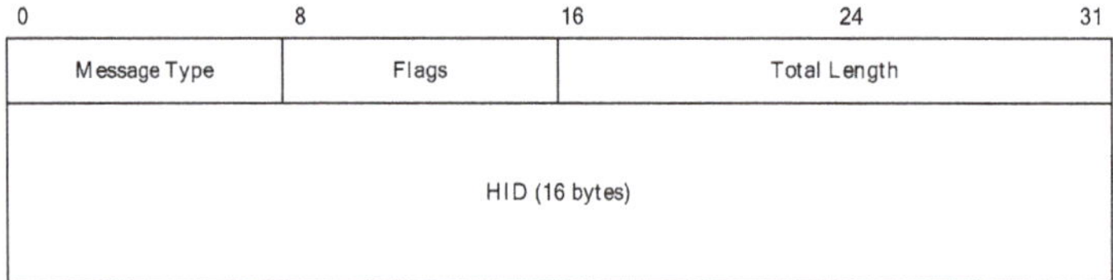

Figure 12. Common header format (20 bytes).

Each control message has a parameter, depending on the case. If there is a parameter in the control message, its type-length-value (TLV) format is as shown in Figure 13.

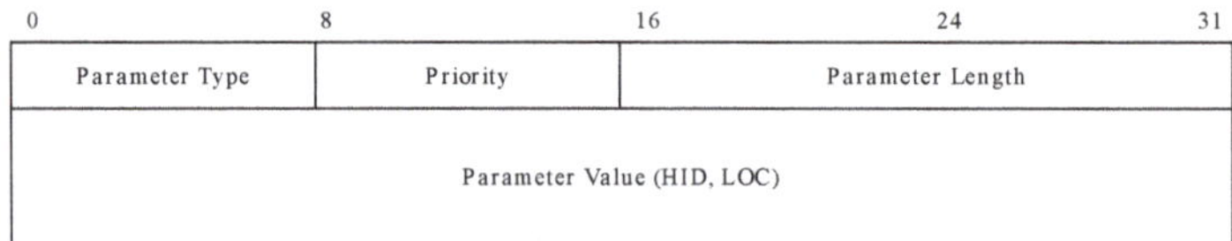

Figure 13. Parameter type-length-value (TLV) format.

In the figure, the parameter type field is an encoding value of the parameter that has either an HID (0000 0000) or LOC (0000 0001). The priority field indicates the priority of this parameter when two or more parameters are contained. Specifically, "0" represents that no priority is given, and "1" represents the highest priority, whereas "255" is the lowest one. The parameter length field is the length of this parameter in bytes. Finally, the parameter value has an assigned variable size for the HID or LOC.

4. MOFI Implementation

We implemented the MOFI architecture using OpenFlow [16] and Click Modular Router [17] over a Linux platform. Next, we describe implementation details.

4.1. Host

Firstly, we set up two hosts. One was a personal computer (PC)-based host and the other was an Android-based host. Due to the page limitations, however, we only describe the PC-based host in this article. Figure 14 shows a protocol stack and its connection with other entities.

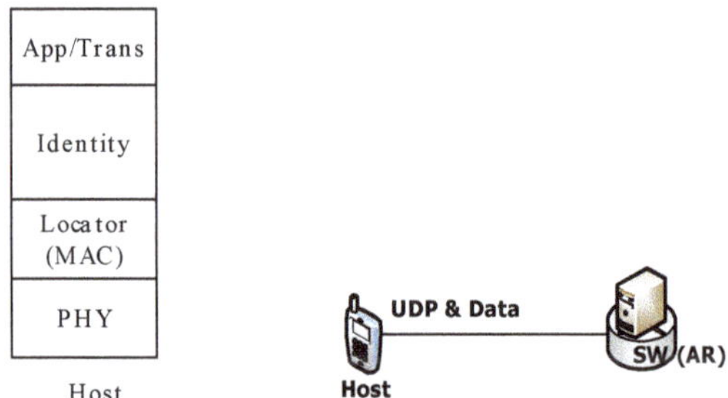

Figure 14. Protocol stack of host and connection with other entities.

In the figure, the host reuses the IPv6 protocol stack as an identity stack for backward compatibility. Thus, it is possible to use traditional socket interfaces for application programming. The locator field is used for packet delivery between a host and SW, which is used by a traditional MAC protocol stack. The host exchanges the data packet and HID binding operation with the SW. We implemented the HBR and HBA messages using raw socket APIs for the Internet Control Message Protocol version 6 (ICMPv6) protocol.

4.2. Switch

For the implementation of the SW, we used OpenFlow that was built on top of OpenVSwitch. Figure 15 shows the protocol stack of the SW and its connection with other entities.

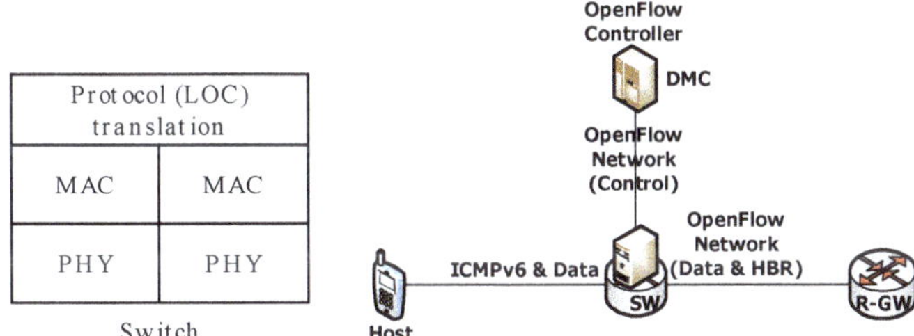

Figure 15. Protocol stack of the switch (SW) and connection with other entities.

SW translates a protocol (LOC) by changing the MAC header and uses an OpenFlow network. The SW exchanges the HID binding update operation and sends data packets to the R-GW through the OpenFlow network and performs the LOC query operation for the DMC through the OpenFlow control channel. In the OpenFlow network, the LOC query operation is replaced by the packet-in and packet-out messages.

4.3. R-GW

For the implementation of the R-GW, we used OpenFlow and Click Modular Router. Figure 16 shows the protocol stack of the R-GW and its connection with other entities.

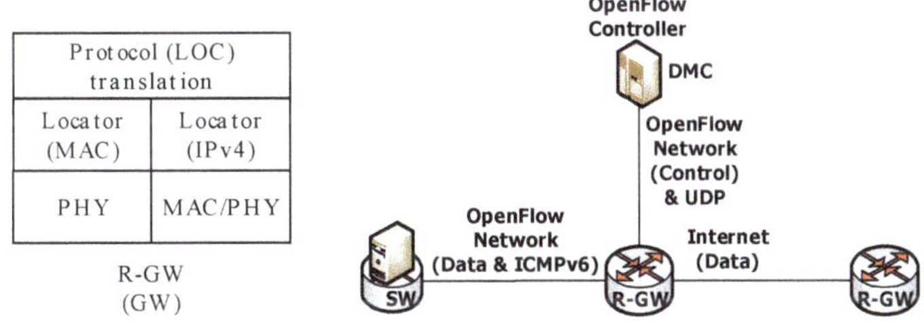

Figure 16. Protocol stack of the regional gateway (R-GW) and connection with other entities.

The R-GW translates the protocol (LOC) by changing the MAC and an IPv4 headers and uses an OpenFlow network with the SW and DMC. The R-GW exchanges the HID binding update operation, sends data packets to the SW through the OpenFlow network, and performs the LOC query operation and HID binding update operation for the DMC through the UDP socket API. We used Click Modular Router to translate the LOC header of data packets, as well as to encapsulate and de-capsulate the HID binding update and LOC query message to a UDP packet.

Figure 17 shows how a packet of the Click Modular Router at the R-GW is processed. In this figure, the R-GW has three network interfaces named INT I/F, Control I/F, and EXT I/F. INT I/F stands for an internal interface, which is connected to SWs via an OpenFlow data channel using an IPv6 network. Control I/F stands for a control interface, which is connected to the DMC through the

OpenFlow control channel using the IPv4 network. Finally, EXT I/F stands for an external interface, which is connected to other R-GWs through the Internet. Upon receiving a packet from the SW, the INT I/F forwards it to the `Classifier()`, which classifies the packet to an IPv6 Neighbor Discovery Solicitation (IPv6 NDS), IPv6 Neighbor Advertisement (IPv6 NDA), IPv4, or IPv6 packet. An IPv6 NDS packet is forwarded to the IP6NDadvertisement element, while an IPv6 NDA packet is forwarded to the IP6NDSolicitor element. Then, they return to the INT I/F. An IPv4 packet is stripped to the MAC header and IPv4 header by the `Strip()`. After that, Click Modular Router adds an IP and a UDP header through `UDPIPEncap()`. To construct and encapsulate a MAC header, the packet is sent to `ARPQuerier()` and the DMC through the Control I/F.

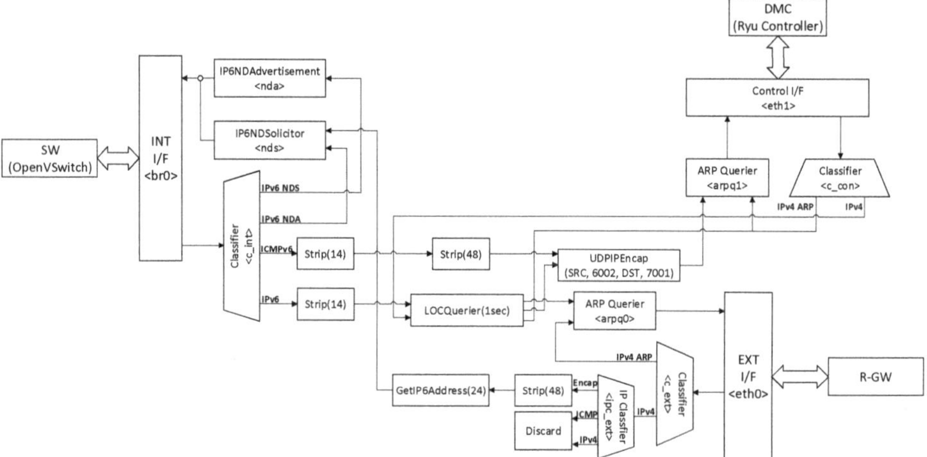

Figure 17. Processing a packet at the R-GW using Click Modular Router.

An IPv6 packet is also stripped to the MAC header by `Strip()`. After that, if it is a data packet, it is forwarded to `LOCQuerier()` that is implemented for performing the LOC query operation and encapsulating the packet to IPv4 to forward another R-GW through the Internet. `LOCQuerier()` searches the DFC to find the LOC of RH. If there is no RH LOC at the DFC, it performs the LOC query operation for the DMC. To perform the LOC query operation, `LOCQuerier()` makes the LQR message and forwards it to `UDPIPEncap()` to construct the IP and UDP header. Then, the message is forwarded to the DMC through the Control I/F. DMC processes the LQR message and relays the LQA message to the R-GW through the Control I/F. The Control I/F forwards all receiving packets to the `Classifier()`, which classifies the packet into IPv4 ARP and others. The ARP packet is forwarded to `ARPQuerier()`. An LQA message is forwarded to `LOCQuerier()`. When receiving an LQA message, `LOCQuerier()` records the RH's HID and the RH's LOC to the DMC. At the same time, it constructs and encapsulates a MAC header. If the RH's LOC exists at the DFC, the LOC query operation is omitted and `LOCQuerier()` is performed to construct and encapsulate a data packet. Then, the constructed packet is sent to `ARPQuerier()` and to another R-GW through the EXT I/F. Contrariwise, receiving a packet from another R-GW, EXT I/F forwards it to the `Classifier()`, which classifies the packet as IPv4 ARP and others. The ARP packet is forwarded to `ARPQuerier()`. Another packet is classified by the `IPClassifier()` into data, or ICMPv6 or IPv6 packets. The data packet is stripped to the encapsulated header at `Strip()`. Then, through `GetIP6Address()`, its IPv6 header is marked. To construct and encapsulate an MAC header, the packet is sent to the `IP6DNSolicitor()` element and it is sent to the SW through the INT I/F.

In the R-GW, we implemented `LOCQuerier()`, which maintains its DFC using cache memory, performs an LOC query operation, and encapsulates data packets. Figure 18 shows `LOCQuerier()`'s architecture.

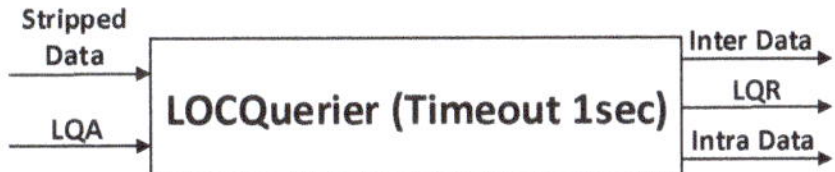

Figure 18. `LOCQuerier()` element's architecture.

To implement `LOCQuerier()`, we referred the `ARPQuerier()` element that handles all the data packets arriving at the R-GW. The argument timeout should be the timer of the DFC cache expiration. There are two input ports and three output ports. Packets arriving at input 0 should be stripped data packets and must have a destination HID. If a DFC cache for the destination HID already exists, the data packets are sent to the inter-domain or intra-domain in accordance with their destination HID. If a DFC cache does not exist, data packets are saved and an LQR message is sent instead. Then, an LQA message should include the LOC of the destination. At the same time, a DFC cache is created, and it saves HID–LOC mapping information. The DFC cache is automatically deleted after one second because of the expiration timeout.

For inter-domain communications, a data packet is encapsulated by `LOCQuerier()` into an IPv4 header and is encapsulated by `ARPQuerier()` into a MAC header.

4.4. DMC

For the implementation of the DMC, we used OpenFlow, which uses a Ryu controller because of the IPv6 support. Figure 19 shows the connection of the DMC with other entities.

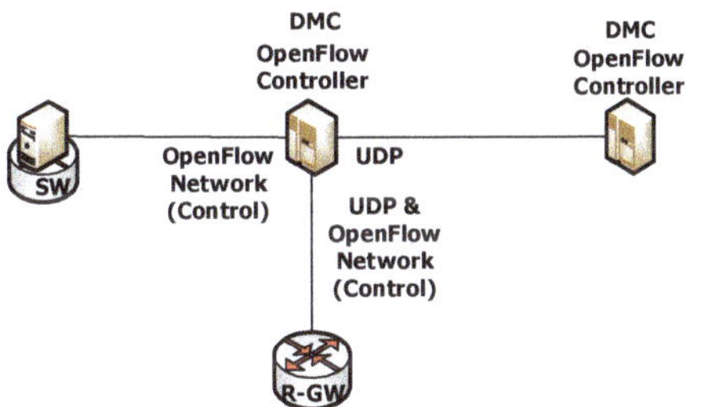

Figure 19. Connection of the distributed mobility controller (DMC) with other entities.

DMC exchanges an HID binding update and an LOC query operation to the R-GW using UDP through an OpenFlow control channel and an LOC query operation to the SW through an OpenFlow control channel, which is located in the intra-domain. Within the DMC, they exchange an HID binding update and an LOC query operation through the UDP. If the DMC receives an HBR/LQR, it determines the destination of the packet, which may be itself or another DMC. Then, the HBR/LQR message is forwarded to the determined destination. In the HBR message's case, the DMC updates its SMR table and creates an HBA message to notify the result of the HBR message. In the case of an LQR message, the DMC searches an SMR table to find the HID–LOC mapping information and creates an LQA message containing the HID–LOC mapping information. At that time, the DMC sends the LQA messages in response to the LQR message.

In MOFI, we assigned OpenFlow entities to the MOFI register and cache because they perform the same role in a network. Table 4 summarizes the entity mapping between OpenFlow and the MOFI implementation.

Table 4. Entity mapping between OpenFlow and the MOFI implementation.

OpenFlow	MOFI Implementation
Flow table (OpenVSwitch)	DFC@SW
Flow table (Ryu controller)	LMR
Packet-in/out/Flow Mod	LQR/LQA (Intra-domain)

The flow table of the Ryu controller is mapped to the DMC's LMR and the flow table of OpenVSwitch is mapped to the DFC of the SW. In the OpenFlow network, the packet-in, packet-out, and flow-mod message set up the route to deliver a data path and, hence, we use these messages to perform the LOC query operation in the MOFI architecture. Through this process, we can reduce the consumption of network resources by avoiding duplicated operations, as well as by reducing the programming burden.

5. MOFI Experimentations on EU–Korea Testbed

The interconnection of Europe and South Korea takes advantage of the GÉANT [19] network in Europe, while the respective KOREN [20] and KREONET [21] are used in South Korea. The GÉANT and KOREN/KREONET are interconnected via TEIN3 [22] and TEIN4 [23].

The first pilot evaluated in this paper was an identity-based communication. Physically distributed sensors across European countries continuously generate data and then they are streamed to South Korea. Because sensors are highly mobile (e.g., portable sensors attached to moving hosts), they cannot be identified using the traditional IP addressing mechanism. The communications between Europe and South Korea sites are achieved using a host identifier, while IP addresses are only used in an inter-domain area. When a host enters the range of a new switch (SW), the sensor data are forwarded to the new SW by a location management function operated by OpenFlow controller. Each host is able to start a sensor data streaming service with every host, whether located in Europe or South Korea, where the sensor data can be forwarded constantly and seamlessly. This service is orchestrated by another OpenFlow controller. In this scenario, the OpenFlow controller is named a distributed mobility controller (DMC).

Another pilot was a content-based communication. In this scenario, a content-based architecture was implemented using SDN technologies on top of Europe and South Korea testbeds. The utilized resources were interconnected including Layer 2 intercontinental virtual links, based on GEANT–GLORAID–KREONET services. Wireless devices laying in the Europe testbed are connected to a content-based network in South Korea, and content identifiers are used instead of IPs. The goal of this innovation is to use identifiers only to specify the content itself, unlike an IP address which specifies the location of a content. Each content is placed on multiple sides of the South Korea testbed. The target of the content-based architecture follows the content from the most appropriate side to the requesting wireless device, while the streaming over the Europe wireless mesh is based on a backpressure routing scheme.

In the following section, we discuss the identity-based communication scenario that uses the MOFI architecture in detail. Then, we evaluate the MOFI architecture by constructing the testbeds in Europe and South Korea.

5.1. Testbed Configuration

In the discussion below, we describe the experiments conducted over the testbed.

Figure 20 shows testbed configurations for the experiments. There were three network domains as follows: one was located at KNU (Kyungpook National University), another was located at

ETRI (Electronics and Telecommunications Research Institute), and the last one was located at UMU (Universidad de Murcia). On the Korea side, KNU and ETRI sites are connected via KOREN [20] and KREONET [21]. On the EU side, UMU uses the GAIA network. Korea and the EU are connected via TEIN4 [23]. For the experiments, we used Ubuntu 12.04 and Linux kernel 3.5.0.23-generic version. To support the MOFI HID, we used OpenFlow 1.3 version, OpenVswitch, and Ryu controller. Each domain has a unique AS (Autonomous System) number. KNU was assigned to 0×2744, and ETRI was assigned to $0 \times 0EA4$. However, because the UMU site was not assigned yet, it used ETRI's AS number temporarily.

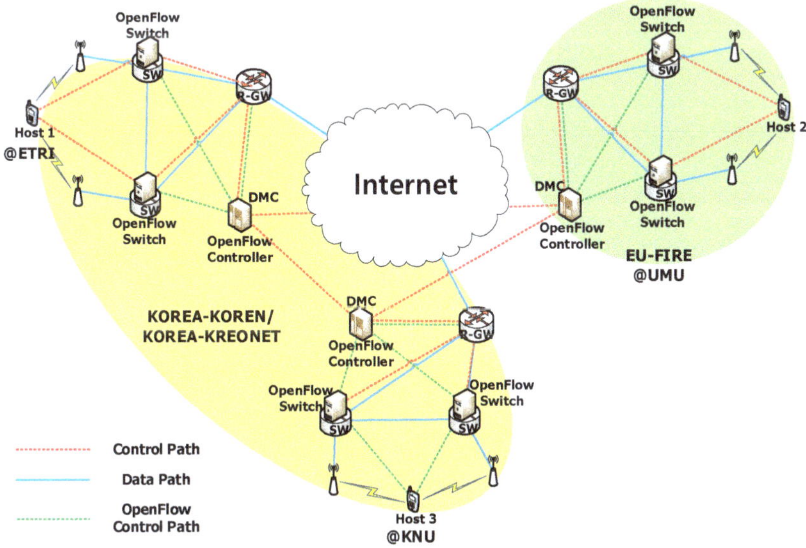

Figure 20. Testbed configuration for the MOFI evaluation.

Each domain had the same intra-domain network architecture except for hardware specifications, as shown in Figure 21. The host had a globally unique HID that was connected to the SW through an IPv6 network to deliver data packets and HID binding messages. To support a wireless network, we used an access point, and the SW also supported multiple hosts. The SW used OpenVSwitch and had three interfaces. The first interface was the OpenFlow controller of this domain, which was connected to the DMC. It used an IPv4 network to transmit an OpenFlow control message instead of an LOC query operation, in which the interface used a 192.168.1.2 IP address. Otherwise, the other two interfaces used an IPv6 network to forward data packets and HID binding messages to other entities. These interfaces did not have an IPv6 address because they configured the OpenFlow network. One interface was connected to the host, while another one was connected to the R-GW to communicate between other domain hosts and to perform an HID binding operation, which used a MAC header as the LOC. The R-GW used OpenVSwitch and Click Modular Router and also had three interfaces. One was connected to the DMC and used an IPv4 network to transmit OpenFlow control messages, HID binding messages, and LOC query messages. It used an IP address of 192.168.1.3. To perform an HID binding operation and an LOC query operation, it used a Click Modular router to translate HID binding messages and to send LOC query messages. The other interface used an IPv6 network to transmit the data packets and HID binding messages to the SW, the interface of which had an IPv6 address of 2020::99:99:99:99 as a default gateway. The last interface used a public IPv4 network to transmit data packets to other domains. These communications could be achieved by the LOC translation using the Click Modular router. The DMC had two interfaces. One was connected to the SW and R-GW to transmit OpenFlow control messages, HID binding messages, and LOC query messages. The other was connected to other DMCs belonging in a different domain. Thus, it used a public IPv4 network. To transmit HID binding messages and LOC query messages, the DMC used a UDP socket.

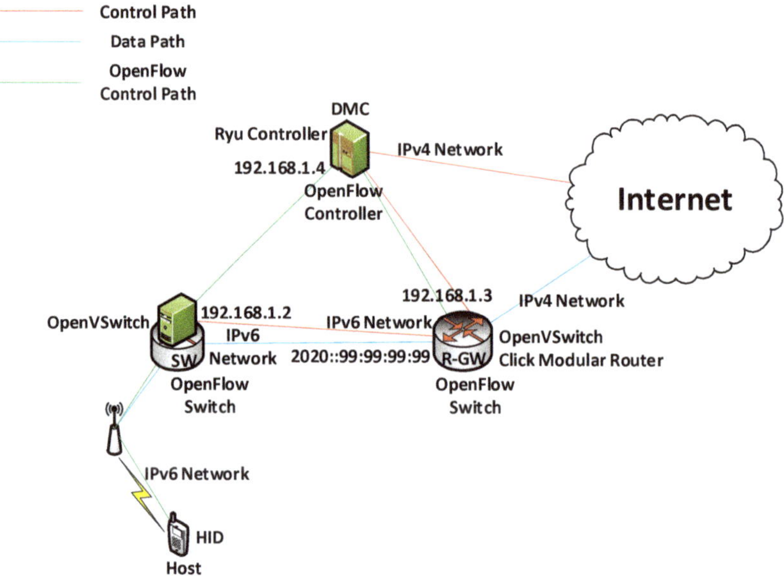

Figure 21. Intra-domain network architecture.

To evaluate our MOFI implementation, each domain had different testbed configurations. KNU and UMU sites consisted of PCs as the SW, R-GW, and DMC. Otherwise, the ETRI site consisted of blade servers to construct the SW, R-GW, and DMC. The hosts located at KNU and ETRI used a laptop-based host. On the other hand, UMU used a PC-based host. In addition, ETRI use smartphone-, tablet-, and TV-based hosts to demonstrate an N-Screen scenario.

5.2. Validation of MOFI Implementations

For the validation of MOFI, we constructed a small testbed locally located at KNU and experimented on the implemented MOFI architecture. Some test scenarios were used for the evaluation. Firstly, Host 1 and Host 2 were attached to each domain. Two applications were used for the validation. One was a UDP echo server/client. Another was a video streaming service. In the UDP echo server/client, Host 1 was a UDP echo server and Host 2 was a UDP echo client. For the streaming service, Host 2 begins receiving the video data traffic from Host 1. To validate the MOFI implementation, we captured the data and control packets that flew in the testbed network using Wireshark [24]. Figure 22 shows the testbed configuration for validation. There were two domains inter-connected by the KOREN backbone network.

In the testbed network, Host 1 was a video streaming server and Host 2 was a VLC player [25]. Host 1 and 2 used 2014::11:11:11:11 and 2014::22:22:22:22 for their HIDs, respectively. As for the LOC, an MAC address was used for the intra-domain communication. On the other hand, a public IPv4 address (155.230.23.183 and 155.230.23.186) was used as the LOC for inter-domain communications. For an OpenFlow control path, we used a private IPv4 address (192.168.1.x). On the other hand, the control path between DMCs used public IPv4 addresses (155.230.23.184 and 155.230.23.185). In the figure, the control path represents the route used for HID–LOC binding and LOC update operations in the MOFI architecture. Figure 23 shows a snapshot of the testbed described in Figure 22.

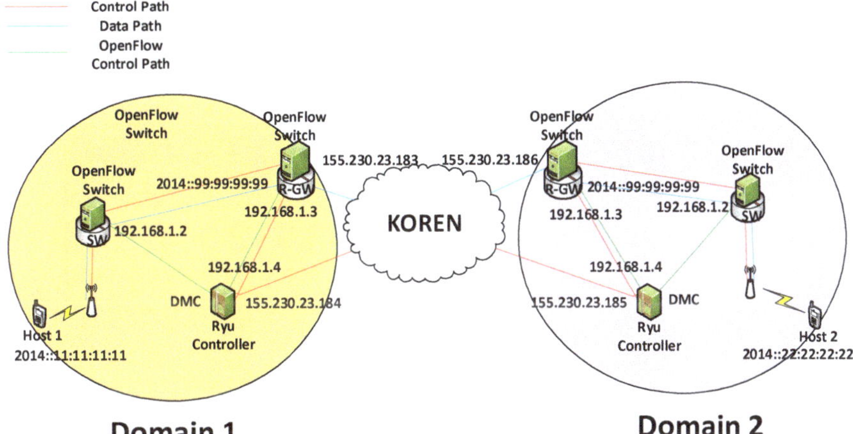

Figure 22. Validation testbed network configuration.

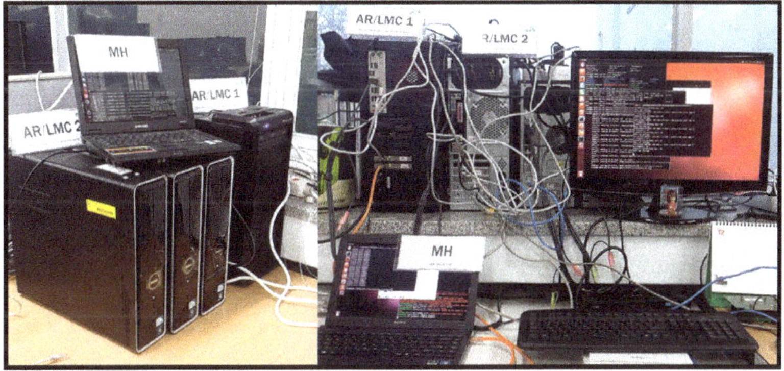

Figure 23. Testbed snapshot for validation.

With this testbed, we validated our MOFI implementation. In the figure, the left side is domain 1 and the right side is domain 2. They had similar network configurations with one mobile host (MH), SW, R-GW, and DMC in each domain. R-GWs and DMCs were interconnected by the KOREN network for inter-domain communications. Each domain was an OpenFlow network.

Figure 24 shows packet capturing results flowing from Host 2 at domain 2. In this figure, Host 1 and 2 use 2014::11:11:11:11 and 2014::22:22:22:22. The HID header is represented as an IPv6 header because Wireshark does not support the MOFI architecture. However, we can know that the MOFI HID header and IPv6 header are interoperable between each other at the application layer, and they use a similar packet format. Moreover, we can know that the application program is compatible with both the MOFI architecture and an IPv6 network.

Figure 25 shows the packet capturing result flowing from the R-GW of domain 1 to the R-GW of domain 2. In this figure, we can see that LOCs were translated along the path since the Click Modular Router of R-GW performed the protocol translation by encapsulating the LOC header. While MAC and IPv6 headers were used as LOC and HID, respectively, R-GWs used 155.230.23.186 and 155.230.23.183 as LOCs (public IP addresses). Furthermore, the packet size was larger than the Pv4 header size (20 bytes) because R-GW encapsulated an IPv4 header into an LOC header. We can also see that a MAC header was successfully translated. In the meantime, HIDs of Host 1 and Host 2 did not change during the data delivery.

No.	Time	Source	Destination	Protocol	Length	Info
1 0.000000	2014::22:22:22:22	2014::11:11:11:11	UDP	67	Source port: 46297 Destination port
3 1.000215	2014::22:22:22:22	2014::11:11:11:11	UDP	67	Source port: 46297 Destination port
5 2.000399	2014::22:22:22:22	2014::11:11:11:11	UDP	67	Source port: 46297 Destination port
7 3.000632	2014::22:22:22:22	2014::11:11:11:11	UDP	67	Source port: 46297 Destination port
9 4.000851	2014::22:22:22:22	2014::11:11:11:11	UDP	67	Source port: 46297 Destination port
11 5.001073	2014::22:22:22:22	2014::11:11:11:11	UDP	67	Source port: 46297 Destination port
13 6.001314	2014::22:22:22:22	2014::11:11:11:11	UDP	67	Source port: 46297 Destination port
15 7.001502	2014::22:22:22:22	2014::11:11:11:11	UDP	67	Source port: 46297 Destination port
17 8.001742	2014::22:22:22:22	2014::11:11:11:11	UDP	67	Source port: 46297 Destination port
19 9.001946	2014::22:22:22:22	2014::11:11:11:11	UDP	67	Source port: 46297 Destination port
21 10.002153	2014::22:22:22:22	2014::11:11:11:11	UDP	67	Source port: 46297 Destination port

```
▶ Frame 1: 67 bytes on wire (536 bits), 67 bytes captured (536 bits)
▶ Ethernet II, Src: Sony_87:4c:f2 (54:42:49:87:4c:f2), Dst: EfmNetwo_44:6e:4f (00:26:66:44:6e:4f)
▶ Internet Protocol Version 6, Src: 2014::22:22:22:22 (2014::22:22:22:22), Dst: 2014::11:11:11:11 (2014::11:11:11:11
▶ User Datagram Protocol, Src Port: 46297 (46297), Dst Port: 8854 (8854)
▶ Data (5 bytes)
```

Figure 24. Packet capture (at host 2).

No.	Time	Source	Destination	Protocol	Length	Info
48 0.933938	2014::22:22:22:22	2014::11:11:11:11	UDP	87	Source port: 46297 Destination port: 8854
83 1.934157	2014::22:22:22:22	2014::11:11:11:11	UDP	87	Source port: 46297 Destination port: 8854
122 2.934380	2014::22:22:22:22	2014::11:11:11:11	UDP	87	Source port: 46297 Destination port: 8854
167 3.934613	2014::22:22:22:22	2014::11:11:11:11	UDP	87	Source port: 46297 Destination port: 8854
205 4.934833	2014::22:22:22:22	2014::11:11:11:11	UDP	87	Source port: 46297 Destination port: 8854
238 5.935037	2014::22:22:22:22	2014::11:11:11:11	UDP	87	Source port: 46297 Destination port: 8854
279 6.935278	2014::22:22:22:22	2014::11:11:11:11	UDP	87	Source port: 46297 Destination port: 8854
331 7.935432	2014::22:22:22:22	2014::11:11:11:11	UDP	87	Source port: 46297 Destination port: 8854
379 8.935602	2014::22:22:22:22	2014::11:11:11:11	UDP	87	Source port: 46297 Destination port: 8854
426 9.935859	2014::22:22:22:22	2014::11:11:11:11	UDP	87	Source port: 46297 Destination port: 8854
466 10.936024	2014::22:22:22:22	2014::11:11:11:11	UDP	87	Source port: 46297 Destination port: 8854
514 11.936265	2014::22:22:22:22	2014::11:11:11:11	UDP	87	Source port: 46297 Destination port: 8854
566 12.936468	2014::22:22:22:22	2014::11:11:11:11	UDP	87	Source port: 46297 Destination port: 8854

```
▶ Frame 48: 87 bytes on wire (696 bits), 87 bytes captured (696 bits)
▶ Ethernet II, Src: MarvellS_00:10:c1 (00:50:43:00:10:c1), Dst: EdimaxTe_fa:50:3f (00:1f:1f:fa:50:3f)
▶ Internet Protocol Version 4, Src: 155.230.23.186 (155.230.23.186), Dst: 155.230.23.183 (155.230.23.183)
▶ Internet Protocol Version 6, Src: 2014::22:22:22:22 (2014::22:22:22:22), Dst: 2014::11:11:11:11 (2014::11:11:11:11)
▶ User Datagram Protocol, Src Port: 46297 (46297), Dst Port: 8854 (8854)
▶ Data (5 bytes)
```

Figure 25. Packet capture (at the R-GW of domain 1).

To validate the control plane operation, we used Wireshark to capture control packets at the R-GW and DMC. Figure 26 shows the packet capturing result for the LOC query operations between the R-GW and DMC.

No.	Time	Source	Destination	Protocol	Length	Info
1 0.000000	192.168.1.3	192.168.1.4	UDP	138	Source port: x1
2 1.000193	192.168.1.3	192.168.1.4	UDP	138	Source port: x1
3 2.000385	192.168.1.3	192.168.1.4	UDP	138	Source port: x1
6 3.000600	192.168.1.3	192.168.1.4	UDP	138	Source port: x1
7 4.000794	192.168.1.3	192.168.1.4	UDP	138	Source port: x1
8 5.000988	192.168.1.3	192.168.1.4	UDP	138	Source port: x1
11 6.001180	192.168.1.3	192.168.1.4	UDP	138	Source port: x1
12 7.001380	192.168.1.3	192.168.1.4	UDP	138	Source port: x1
13 8.001567	192.168.1.3	192.168.1.4	UDP	138	Source port: x1
14 9.001782	192.168.1.3	192.168.1.4	UDP	138	Source port: x1
15 10.001954	192.168.1.3	192.168.1.4	UDP	138	Source port: x1
16 11.002153	192.168.1.3	192.168.1.4	UDP	138	Source port: x1
17 12.002345	192.168.1.3	192.168.1.4	UDP	138	Source port: x1

```
▶ Frame 1: 138 bytes on wire (1104 bits), 138 bytes captured (1104 bits)
▶ Ethernet II, Src: Dell_00:8c:15 (18:03:73:00:8c:15), Dst: Dell_e0:e4:89 (b8:ac:6f:e0:e4:89)
▶ Internet Protocol Version 4, Src: 192.168.1.3 (192.168.1.3), Dst: 192.168.1.4 (192.168.1.4)
▶ User Datagram Protocol, Src Port: x11-2 (6002), Dst Port: afs3-callback (7001)
▶ Data (96 bytes)
```

Figure 26. Packet capture (between the R-GW and DMC).

In the figure, we can see that the source address and port number was 192.168.1.3:6002, and the destination address and port number was 192.168.1.4:7001. Then, the message used UDP. This means that the messages were generated by the Click Modular Router of the R-GW for the LOC query operation, while the OpenFlow control messages used transmission control protocol (TCP). The packet involved LQR messages to query the LOC of the RH to the designated DMC of the RH.

Figure 27 shows the packet capturing result between DMCs. From the figure, a packet is transmitted from the DMC of domain 2 (155.230.23.185) to the DMC of domain 1 (155.230.23.184). We realized that all the packets were LQA messages because Host 1 sent data packets to Host 2 in the test scenario and, thus, the LOC query operation was performed for the DMC of domain 1 to the DMC of domain 2. Moreover, because LQA messages had four parameters to store the HID and LOC of the SH and RH, they were larger than other control messages. The actual packet size was the biggest when compared with other control packets shown in Figure 26, Figure 28, and Figure 29.

```
No.     Time          Source           Destination        Protocol Length Info
     12 0.857617    155.230.23.185   155.230.23.184      UDP         270 Source port: 50407  Destina
     60 1.857805    155.230.23.185   155.230.23.184      UDP         270 Source port: 50407  Destina
    120 2.858016    155.230.23.185   155.230.23.184      UDP         270 Source port: 50407  Destina
    149 3.858191    155.230.23.185   155.230.23.184      UDP         270 Source port: 50407  Destina
    234 4.858382    155.230.23.185   155.230.23.184      UDP         270 Source port: 50407  Destina
    282 5.858584    155.230.23.185   155.230.23.184      UDP         270 Source port: 50407  Destina
    326 6.858772    155.230.23.185   155.230.23.184      UDP         270 Source port: 50407  Destina
    366 7.858962    155.230.23.185   155.230.23.184      UDP         270 Source port: 50407  Destina
    393 8.859160    155.230.23.185   155.230.23.184      UDP         270 Source port: 50407  Destina
    481 9.859349    155.230.23.185   155.230.23.184      UDP         270 Source port: 50407  Destina
    515 10.859546   155.230.23.185   155.230.23.184      UDP         270 Source port: 50407  Destina
    548 11.859738   155.230.23.185   155.230.23.184      UDP         270 Source port: 50407  Destina
    603 12.859972   155.230.23.185   155.230.23.184      UDP         270 Source port: 50407  Destina
▶ Frame 12: 270 bytes on wire (2160 bits), 270 bytes captured (2160 bits)
▶ Ethernet II, Src: DavicomS_00:01:08 (00:60:6e:00:01:08), Dst: DavicomS_00:02:42 (00:60:6e:00:02:42)
▶ Internet Protocol Version 4, Src: 155.230.23.185 (155.230.23.185), Dst: 155.230.23.184 (155.230.23.184)
▶ User Datagram Protocol, Src Port: 50407 (50407), Dst Port: afs3-callback (7001)
▶ Data (228 bytes)
```

Figure 27. Packet capture (between DMCs).

```
No.     Time          Source              Destination          Protocol Length Info
      1 0.000000    2014::22:22:22:22   2014::99:99:99:99    ICMPv6       158 Private experimentation
      3 1.000147    2014::22:22:22:22   2014::99:99:99:99    ICMPv6       158 Private experimentation
      5 2.000301    2014::22:22:22:22   2014::99:99:99:99    ICMPv6       158 Private experimentation
      7 3.000444    2014::22:22:22:22   2014::99:99:99:99    ICMPv6       158 Private experimentation
      9 4.000589    2014::22:22:22:22   2014::99:99:99:99    ICMPv6       158 Private experimentation
     11 5.000744    2014::22:22:22:22   2014::99:99:99:99    ICMPv6       158 Private experimentation
     12 5.814603    fe80::5642:49ff:fe87:  2014::99:99:99:99 ICMPv6        86 Neighbor Solicitation for 2014::99:99:
     13 5.814639    2014::99:99:99:99   fe80::5642:49ff:fe87:  ICMPv6      78 Neighbor Advertisement 2014::99:99:99:
     14 5.814639    2014::99:99:99:99   fe80::5642:49ff:fe87:  ICMPv6      86 Neighbor Advertisement 2014::99:99:99:
     16 6.000888    2014::22:22:22:22   2014::99:99:99:99    ICMPv6       158 Private experimentation
▶ Frame 5: 158 bytes on wire (1264 bits), 158 bytes captured (1264 bits)
▶ Ethernet II, Src: Sony_87:4c:f2 (54:42:49:87:4c:f2), Dst: Broadcast (ff:ff:ff:ff:ff:ff)
▶ Internet Protocol Version 6, Src: 2014::22:22:22:22 (2014::22:22:22:22), Dst: 2014::99:99:99:99 (2014::99:99:99:99)
▼ Internet Control Message Protocol v6
   Type: Private experimentation (200)
   Code: 0
   Checksum: 0xf882 [correct]
 ▼ [Expert Info (Note/Undecoded): Dissector for ICMPv6 Type (200) code not implemented, Contact Wireshark developers
     [Message: Dissector for ICMPv6 Type (200) code not implemented, Contact Wireshark developers if you want this sup
     [Severity level: Note]
```

Figure 28. Packet capture (at the R-GW of domain 2).

```
No.     Time          Source           Destination        Protocol Length Info
      4 3.000435    192.168.1.3      192.168.1.4         UDP         138 Source port: x1
      5 4.000588    192.168.1.3      192.168.1.4         UDP         138 Source port: x1
      6 5.000736    192.168.1.3      192.168.1.4         UDP         138 Source port: x1
      7 6.000860    192.168.1.3      192.168.1.4         UDP         138 Source port: x1
      8 7.001003    192.168.1.3      192.168.1.4         UDP         138 Source port: x1
      9 8.001126    192.168.1.3      192.168.1.4         UDP         138 Source port: x1
     10 9.001281    192.168.1.3      192.168.1.4         UDP         138 Source port: x1
     11 10.001436   192.168.1.3      192.168.1.4         UDP         138 Source port: x1
     12 11.001610   192.168.1.3      192.168.1.4         UDP         138 Source port: x1
     13 12.001753   192.168.1.3      192.168.1.4         UDP         138 Source port: x1
▶ Frame 1: 138 bytes on wire (1104 bits), 138 bytes captured (1104 bits)
▶ Ethernet II, Src: Dell_00:8c:15 (18:03:73:00:8c:15), Dst: Dell_e0:e4:89 (b8:ac:6f:e0:e4:89)
▶ Internet Protocol Version 4, Src: 192.168.1.3 (192.168.1.3), Dst: 192.168.1.4 (192.168.1.4)
▶ User Datagram Protocol, Src Port: x11-2 (6002), Dst Port: afs3-callback (7001)
▶ Data (96 bytes)
```

Figure 29. Packet capture (at R-GW of domain 1).

Figure 28 shows the packet capturing result for the HID binding operation of Host 2. In the figure, we can see that the destination address was 2014::99:99:99:99. This was because the host could not know the destination address to perform the HID binding update operation when attached to a new network. Thus, we supposed that all R-GWs should have a bridge interface (br0) to be the destination of the HID binding update message, for which the address was 2014::99:99:99:99. To perform the HID binding operation, we implemented an Internet Control Message Protocol version 6 (ICMPv6) message using raw socket APIs, because ICMPv6 has several options such as the neighbor discovery protocol (NDP). Thus, we concluded the ICMPv6 message to perform the HID binding operation. An ICMPv6 message was extended to perform the HID binding operation between a host and the R-GW. For this purpose, we assigned the type as 200, which means private experimentation, and we implemented an ICMP message body as an HBR/HBA message. Because of using ICMPv6 and supporting the backward compatibility between HID and IPv6, we were able to implement the MOFI architecture while still allowing the use of the existing IPv6 network and IPv6 applications.

Figure 29 shows the packet capturing result between the R-GW and the DMC for the HID binding operation of Host 1. From the figure, we can see the source address and its port. Then, the messages used UDP for the transmission. Unlike LQR messages, these messages were not generated by the Click Modular Router of the R-GW. Upon receiving HBR messages from the Host, they were encapsulated by the Click Modular Router of R-GW for the HID binding operation and they were forwarded to their own DMC.

5.3. N-Screen Scenario

We chose an N-Screen application to demonstrate the superiority of the MOFI architecture since this service scenario occurs frequently and it shows the service mobility scenario. When a user returns home from outside, the video still streams to the smart phone. At the time a user enters the house, various screens are discovered, for example, television (TV), tablet, etc. These screens perform negotiation processes with each other and share the HID. At this point, the smart phone selects another destination screen that will receive the ongoing video stream. Since the MOFI GW maintains its mapping table that maps each screen's HID to its LOC, the selected screen is assigned the same HID and the video is directly forwarded to the desired screen through the GW.

For this purpose, we implemented the N-Screen application based on an Android system and constructed two domains at the ETRI site. There are two domains that are interconnected by the KREONET backbone network. Furthermore, we used one domain of the KNU site for the server side of the N-Screen scenario; the video stream initiated a server located at KNU and video clients located at ETRI, and we could observe the media data being streamed through the network, which was inter-connected by the EU–Korea network.

Figure 30 shows the testbed for the N-Screen scenario at the ETRI site. There were three screens as follows: the first one was a controller, and the others were screens. There were two screen types for the demonstration. One was a TV using a Universal Serial Bus (USB) dongle based on Android, and the other was a tablet. The controller used the smartphone. For the experiment, Figure 31 shows the controller and two screens.

Firstly, we demonstrated the handover scenario from domain 2 to domain 1. For this scenario, we used the controller as a host device. In this service mobility scenario, the controller was already connected to the MOFI domain 2. Figure 32 shows the inter-domain handover of the controller.

From the figure, the left side shows a screenshot before the handover and the right side shows a screenshot after the handover. We implemented that the user interface of controller can choose one of two access points connected to each domain. In this experiment, we carried out the handover by selecting another access point. Even though we observed handover delay, the handover scenario was successfully performed.

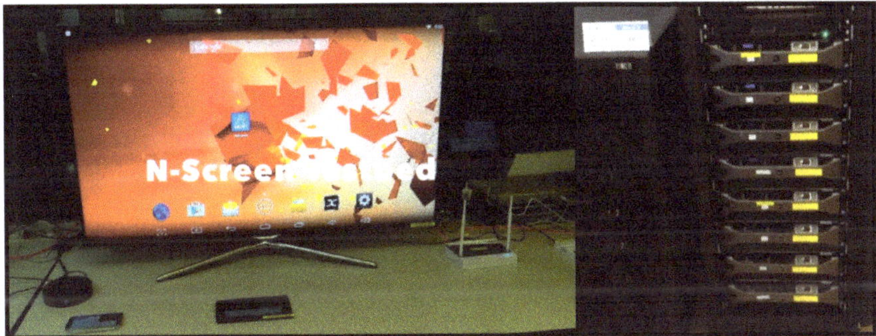

Figure 30. Testbed snapshot for the N-Screen scenario.

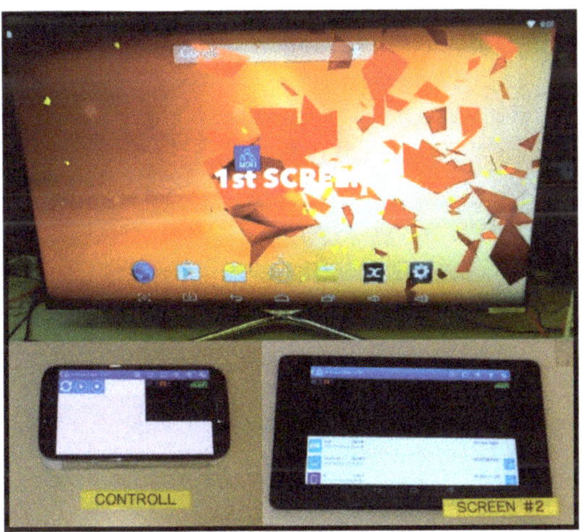

Figure 31. Testbed snapshot for the N-Screen scenario.

Figure 32. Inter-domain handover scenario.

Next, we demonstrated the N-Screen scenario. Using the controller, we selected a screen to play the movie clip. Figure 33 shows screenshots of our N-Screen experiment. The registered screens at the controller are displayed in a list, and a user can then choose one to play. Through the experiment, we could see that a movie clip was properly played on the screen when it was selected. Since screens 1 and 2 were connected to the same domain, it was possible to verify intra-domain handover through this scenario.

Figure 33. N-Screen scenario.

6. Conclusions

In this article, we presented a new mobility management architecture for a future mobile network. The new architecture features the separation of data and control planes, as well as a novel distributed HID–LOC mapping control. We implemented the architecture using OpenFlow and Click Modular Router over a Linux platform and tested the implemented architecture over the EU–Korea testbed network for validation.

To evaluate the proposed architecture, we implemented OpenFlow and Click Modular Router over a Linux platform, and then we validated it using a local testbed. Moreover, we performed the evaluation over an internationally configured EU–Korea testbed network. In particular, we operated the realistic service scenario over the EU–Korea testbed network using an N-Screen scenario. Using various screens for streaming a movie clip, the mobility and the service scenario of the proposed architecture were shown. In the intra-domain mobility event (changing the screen using a controller), there was no observable impact on the streaming session. On the other hand, in the inter-domain mobility event (moving to another domain), although the LOC changed, the HID was able to communicate constantly, and it could be confirmed that there was slight handover latency due to LOC change. However, since the service uses ID-based communication, there is no need to disconnect the service or make a new connection for the service. In order to provide a seamless streaming service, the MOFI control plane performs mapping of each host's HID to a specific LOC, and it updates this mapping information and creates the flows that will forward the traffic to the new location. After updating the mapping information, streaming data are forwarded to the new domain network to which the client is now attached.

In particular, the proposed architecture could provide mobility management without any modification of the current Internet architecture. Furthermore, we showed that the implemented architecture can support backward compatibilities with current IPv6 applications and Internet Protocol networks.

As a future research direction, we will consider integrating security and mobility functionality into the proposed architecture. Because our first goal was to provide an architecture that functions with basic network features, we focused on designing and evaluating the basic architecture. As future research directions, we will firstly address security and mobility issues that were not fully considered in the current design. Also, we will evaluate our architecture with various user scenarios, because Internet services are becoming more dynamic and diverse. Finally, we plan to integrate the concept of virtualization in the next MOFI architecture.

Author Contributions: J.-I.K. wrote the initial manuscript; N.-J.C. and T.-W.Y. performed the testbed experimentation; H.J. conducted the performance analysis; Y.-W.K. and S.-J.K. revised the manuscript.

Acknowledgments: This research was supported by Basic Science Research Program through the National Research Foundation of MoE (NRF-2017R1D1A3B03032156).

References

1. Feldmann, A. Internet Clean-Slate Design: What and Why? *ACM SIGCOMM Comput. Commun. Rev.* **2007**, *37*, 59–64. [CrossRef]
2. Kim, J.I.; Jung, H.; Koh, S.J. Mobile Oriented Future Internet (MOFI): Architectural Design and Implementations. *ETRI J.* **2013**, *35*, 666–676. [CrossRef]
3. Network Simulator 3. Available online: https://www.nsnam.org/ (accessed on 18 March 2019).
4. OPNET. Available online: http://www.riverbed.com/products/performance-management-control/opnet.html (accessed on 18 March 2019).
5. OpenFlow. Available online: http://flowgrammable.org/sdn/openflow/ (accessed on 18 March 2019).
6. Kohler, E.; Robert, M.; Chen, B.; Jannotti, J.; Kaashoek, M.F. The Click modular router. *ACM Trans. Comput. Syst.* **2000**, *18*, 263–297. [CrossRef]
7. Perkins, C.; Johnson, D.; Arkko, J. *Mobility Support in IPv6*. IETF RFC 6275. Available online: https://tools.ietf.org/html/rfc6275 (accessed on 18 March 2019).
8. Gundavelli, S.; Leung, K.; Devarapalli, V.; Chowdhury, K.; Patil, B. *Proxy Mobile IPv6*. IETF RFC 5213. Available online: https://tools.ietf.org/html/rfc5213 (accessed on 18 March 2019).
9. Correia, L.M.; Abramowicz, H.; Johnsson, M.; Wünstel, K. *Architecture and Design for the Future Internet: 4WARD Project*; Springer Science & Business Media: Berlin, Germany, 2011.
10. Lindgren, A. Efficient content-distribution in a hybrid opportunistic network. In Proceedings of the IEEE Consumer Communications and Networking Conference, Las Vegas, NV, USA, 9–12 January 2011.
11. Ahlgren, B.; D'Ambrosio, M.; Marchisio, M.; Marsh, I.; Dannewitz, C.; Ohlman, B.; Pentikousis, K.; Strandberg, O.; Rembarz, R.; Vercellone, V. Design considerations for a network of information. In Proceedings of the 2008 ACM CoNEXT Conference (CoNEXT'08), Madrid, Spain, 9–12 December 2008.
12. Future Internet Design (FIND). Available online: http://www.nets-find.net (accessed on 18 March 2019).
13. Mobility First: Future Internet Architecture. Available online: http://mobilityfirst.winlab.rutgers.edu/ (accessed on 18 March 2019).
14. Raychaudhuri, D.; Nagaraja, K.; Venkataramani, A. MobilityFirst: A robust and trustworthy mobility-centric architecture for the future Internet. *ACM SIGCOMM Comput. Commun. Rev.* **2012**, *16*, 2–13. [CrossRef]
15. Seskar, I.; Nagaraja, K.; Nelson, S.; Raychaudhuri, D. MobilityFirst future internet architecture project. In Proceedings of the 7th Asian Internet Engineering Conference (AINTEC'11), Bangkok, Thailand, 9–11 November 2011.
16. Global Environment for Network Innovations (GENI). Available online: https://www.geni.net/ (accessed on 18 March 2019).
17. Zhang, L.; Afanasyev, A.; Burke, J.; Jacobson, V.; Crowley, P.; Papadopoulos, C.; Wang, L.; Zhang, B. Named data networking. *ACM SIGCOMM Comput. Commun. Rev.* **2014**, *44*, 66–73. [CrossRef]
18. Autonomous System (AS) Numbers. Available online: http://www.iana.org/assignments/as-numbers/as-numbers.xhtml (accessed on 18 March 2019).
19. GEANT. Available online: http://www.geant.org (accessed on 18 March 2019).

20. Korea Research and Education Network (KOREN) Testbed. Available online: http://www.koren.kr (accessed on 18 March 2019).
21. Korea Research Environment Open Network (KREONET) Testbed. Available online: http://www.kreonet.net/ (accessed on 18 March 2019).
22. Trans-Eurasia Information Network 3 (TEIN3) Testbed. Available online: http://www.tein3.net/ (accessed on 18 March 2019).
23. Trans-Eurasia Information Network 4 (TEIN4) Testbed. Available online: http://www.tein4.net/ (accessed on 18 March 2019).
24. WireShark. Available online: http://www.wireshark.org (accessed on 18 March 2019).
25. VLC Media Player. Available online: http://www.videolan.org/vlc/ (accessed on 18 March 2019).

Article

Distributed Identifier-Locator Mapping Management in Mobile ILNP Networks

Moneeb Gohar [1,*]**, Jin-Ghoo Choi** [2,*]**, Waleed Ahmed** [3]**, Arif Ur Rahman** [1]**,
Muhammad Muzammal** [1] **and Seok-Joo Koh** [4]

[1] Department of Computer Science, Bahria University, Islamabad 44000, Pakistan;
     badwanpk@gmail.com (A.U.R.); muzammal.muhammad@gmail.com (M.M.)
[2] Department of Information and Communication Engineering, Yeungnam University,
     Gyeongsan 38541, Korea
[3] Department of Electrical Engineering, University of Poonch, Rawalkot 12350, Pakistan;
     drwaleedahmad@upr.edu.pk
[4] Department of Computer Science and Engineering, Kyungpook National University, Daegu 41566, Korea;
     sjkoh@knu.ac.kr
*   Correspondence: moneebgohar@gmail.com (M.G.); jchoi@yu.ac.kr (J.-G.C.)

Received: 30 October 2019; Accepted: 23 December 2019; Published: 31 December 2019

Abstract: In the Identifier Locator Network Protocol (ILNP) networks, the existing mobility control schemes based on the centralized entity, called the Dynamic Domain Name Service (DDNS) server, such that all the control traffic is processed at the DDNS server. However, the centralized mobility schemes have significant limitations, such as control traffic overhead at the server and large handover delay. In order to resolve these issues, we propose a new mobility control scheme for ILNP networks, which manages the identifier-locators (ID-LOCs) in the fully distributed manner. In our scheme, each domain has a dedicated mobile DDNS (m-DDNS) server at the site border router (SBR). The m-DDNS server maintains two databases; i.e., home host register (HHR) and visiting host register (VHR), to support the roaming of mobile hosts. When a mobile host roams into a domain, the m-DDNS server in the visiting domain registers the host's ID-LOC in the VHR and requests the update of HHR to the m-DDNS server in the home domain. Since the m-DDNS servers communicate each other directly, the ID-LOC mappings are managed without involvement of any central entities. We analyzed our proposed mobility scheme via numerical analysis and compared its performance with those of existing schemes. Numerical results showed that our scheme outperforms the existing mobility control schemes substantially in terms of control traffic overhead at the servers, total transmission delay and handover delay.

Keywords: ILNP; mobility management; identifier-locator; mobile network; roaming

1. Introduction

With the advent of smart phones and various smart devices, the number of mobile users on the Internet is increasing explosively. This mobile trend has caused the rapid growth of the routing table size in routers [1,2]. To tackle the routing scalability issue, the Internet Engineering Task Force (IETF) has proposed the Identifier Locator Network Protocol (ILNP) protocol [3,4], which operates based on the address rewriting scheme. Under the ILNP, the identifier (ID) of a host is separated from its locator (LOC). Specifically, in the IPv6 address of 128 bits, the upper 64 bits are used for LOC and the lower 64 bits are used for ID. Then the Dynamic Domain Name System (DDNS) [5] server maps the ID onto the LOC for each mobile host.

For the ILNP-based networks, a mobility control scheme was proposed in [3,6–17], where individual mobile hosts conduct the binding and query operations with the DDNS server. We call the mobility scheme "ILNP-Global" in this paper, since it depends on the centralized DDNS server for ID-LOC mappings. Unfortunately, the centralized mobility control scheme has significant limitations in scalability and performance. Indeed, as the number of mobile hosts increases, the DDNS server suffers from huge control overhead. Moreover, all the centralized schemes are subject to large operational costs and service quality degradation due to the single point of failure problem.

An alternative mobility control scheme was proposed in [4] to enhance the ILNP-Global, where a local mapping table was employed in addition to the central DDNS server. We call this scheme "ILNP-Local" here. In the ILNP-Local, each domain maintains the local mapping table at its site border router (SBR). The SBR performs the ID-LOC mapping, where the locator is a local LOC (LLOC), for the hosts in the serving domain. On the other hand, when a host is not in the domain, SBR refers to the central DDNS server for the host's locator, which is a global LOC (GLOC). This localized approach improves the scalability and performance over the centralized mobility control scheme. However, it still depends on the central DDNS server for inter-domain ID-LOC mappings.

In this paper, we propose a fully distributed ID-LOC management scheme for mobile ILNP networks, which is denoted by "ILNP-Distributed." In the proposed scheme, we use both the GLOC and LLOC for a mobile host as in the ILNP-Local scheme [4]. We place the dedicated mobile DDNS (m-DDNS) server at each SBR to eliminate the centralized DDNS server. The m-DDNS server manages two databases, called Home Host Register (HHR) and Visiting Host Register (VHR), for ID-LOC mappings. When a host is in the home domain, its ID-LLOC is stored in the HHR of the home domain. On the other hand, when the host moves into another domain, its ID-LLOC is registered in the VHR of the visiting domain and the ID-GLOC is set in the HHR of the home domain. Here, it is the SBR of the visiting domain that requests the ID-GLOC update to the m-DDNS server of the home domain.

The rest of this paper is organized as follows. In Section 2, we review the existing mobility control schemes for ILNP networks. In Section 3, we describe our proposed mobility control scheme. We analyze the considered schemes in Section 4 and compare their performances numerically in Section 5. Finally, we conclude the paper in Section 6.

2. Materials and Methods

In this section, we review two previous mobility control schemes for ILNP networks, which are denoted "ILNP-Global" and "ILNP-Local" respectively.

2.1. ILNP-Global

We consider the mobility scenario shown in Figure 1 to discuss the operation of ILNP-Global. MN (bob.SKT.com) and CN (alice.SKT.com) are located in the same domain initially, but MN moves to another domain then. Further, the MN changes the serving access router (AR) by handover in the visiting domain.

In Figure 2, we illustrate the ID-LOC management of the ILNP-Global scheme in cases of non-roaming, roaming and handover [3]. When MN attaches to AR2 in the home domain (non-roaming case), the MN registers its GLOC by sending a LOC binding update message to the centralized DDNS server (Step 1). The DDNS server registers the MN's ID-GLOC mapping, and returns the LOC binding update ACK to the MN (Step 2). When CN has data packets for the MN, it needs the MN's GLOC for packet delivery, so the CN sends a LOC query request to the DDNS server (Step 3), and obtains the MN's GLOC from the associated LOC query response message (Step 4). We recall that the DDNS server keeps the ID-GLOC information of all the hosts in the network. The CN then transmits the packets directly to the MN since GLOC is the MN's global locator in ILNP-Global, as shown in Figure 1.

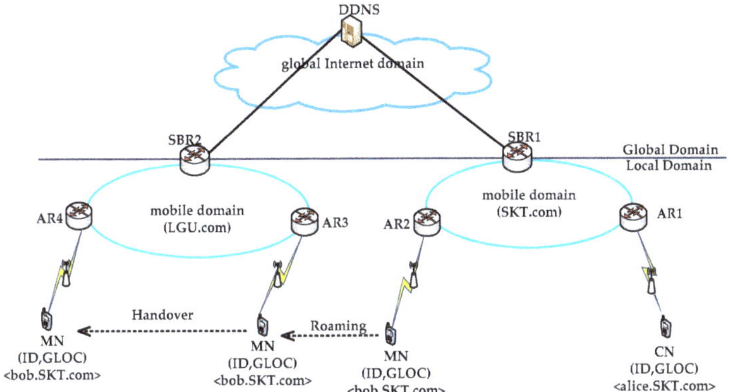

Figure 1. Mobility scenario for Identifier Locator Network Protocol (ILNP)-Global scheme.

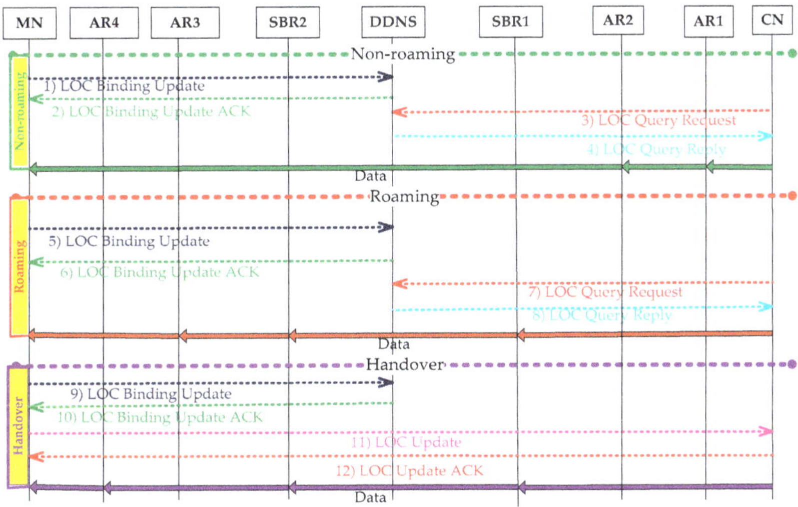

Figure 2. Identifier-locator (ID-LOC) management of ILNP-Global scheme.

When MN moves to another domain and attaches to AR3 (roaming case), the MN updates its GLOC stored in the DDNS server. That is, MN sends a LOC binding update message to the DDNS server (Step 5). The DDNS server updates the MN's ID-GLOC in its database, and returns the LOC binding update ACK to the MN (Step 6). For data delivery, CN requests the MN's GLOC to the DDNS server by sending a LOC query request (Step 7). The DDNS server notifies the MN's GLOC by the LOC query response message (Step 8). The CN can send the data packets to the MN directly.

We now look over the handover of MN from AR3 to AR4. Once MN attaches to the target router AR4, the MN reports its GLOC to the DDNS server. Specifically, MN sends a LOC binding update message with its GLOC to the DDNS server (Step 9). The DDNS server updates the ID-GLOC mapping and returns the LOC binding update ACK to the MN (Step 10). Then, the MN transmits a LOC Update message to CN also, to notify its new GLOC. The CN updates the MN's GLOC, and replies to the MN by the LOC Update ACK (Step 11 and 12). CN can send the data packets to the MN directly with the updated GLOC.

2.2. ILNP-Local

The ILNP-Global scheme depends on the centralized DDNS server to manage the ID-LOC mappings, which incurs significant control message overheads at the server in the global scale. To deal with this problem, the ILNP-Local scheme was proposed in [4]. Differently from ILNP-Global, the ILNP-Local employs a local mapping table (LMT) at the SBR, in conjunction with the DDNS server, to provide the localized mobility control.

For ILNP-Local, we consider the same mobility scenario with that in the ILNP-Global, as shown in Figure 3. MN and CN are located in the same domain but MN moves to another domain by roaming. After some time, the MN makes a handover to another AR in the visiting domain.

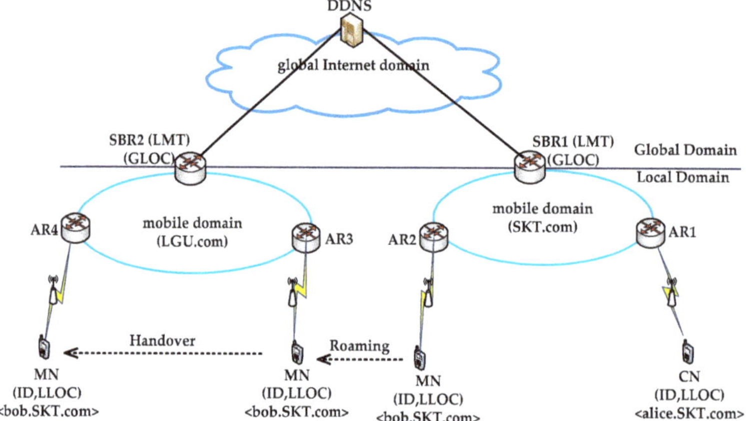

Figure 3. Mobility scenario for ILNP-Local scheme.

In Figure 4, we illustrate the ID-LOC management operations of the ILNP-Local. We first consider the non-roaming case, where MN attaches to AR2 in the home domain. The MN configures its LLOC and registers it to SBR1 by sending a LOC binding update message (Step 1). SBR1 stores the MN's ID-LLOC mapping in its LMT, and responds with the LOC binding update ACK to the MN (Step 2). Subsequently, SBR1 sends a LOC binding update message to the central DDNS server to register the MN's ID-GLOC (Step 3). The DDNS server replies with the LOC binding update ACK to SBR 1 (Step 4) after storing the ID-GLOC in its database. We notice here that the GLOC is the SBR's locator in ILNP-Local, as shown in Figure 3, whereas a GLOC is the MN's locator in ILNP-Global. When CN has data packets for the MN, the packets are delivered to the SBR1 first. Then, SBR1 looks up its LMT to find the MN's LLOC. Since the LLOC is in the LMT in non-roaming case, SBR1 knows the MN's current location and forwards the data packet successfully.

We turn to the roaming case now, where MN attaches to AR3 in the visiting domain. Once the MN is connected to AR3, it configures its LLOC and registers it to SBR2 by sending a LOC binding update message (Step 5). SBR2 stores the MN's ID-LLOC in its LMT and returns the LOC binding update ACK to the MN (Step 6). Right after, SBR2 sends a LOC binding update message to the centralized DDNS server (Step 7) to update the MN's ID-GLOC mapping. The DDNS server then responds with the LOC binding update ACK to SBR2 (Step 8). For data delivery to the MN, CN sends data packets to SBR1 in the home domain. SBR1 searches the LMT but fails to find the MN's LLOC. In that case, SBR1 refers to the DDNS server for the MN's GLOC by sending the LOC query request message (Step 9). When the DDNS server returns the associated LOC query response (Step 10), SBR1 knows the MN's GLOC and forwards the data packets to SBR2 in the visiting domain. The SBR2 delivers the packets to the MN since it has the MN's LLOC in its LMT.

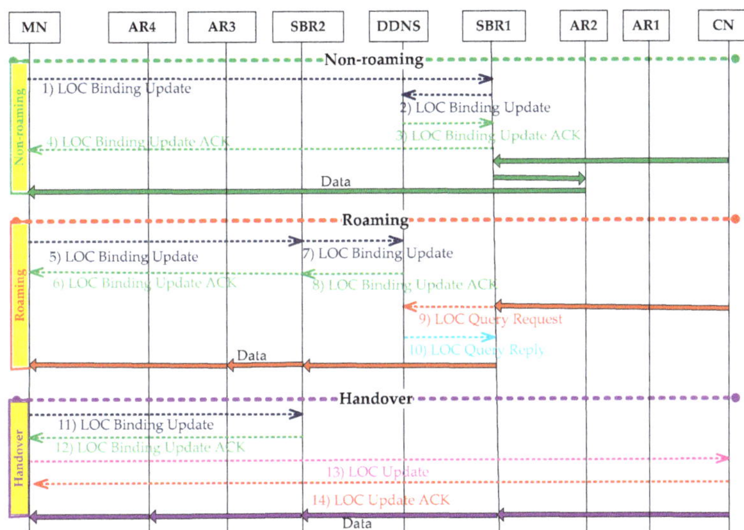

Figure 4. ID-LOC management of ILNP-Local scheme.

During the handover from AR3 to AR4, MN sends a LOC binding update message to SBR2 to update its ID-LLOC in the LMT (Step 11). The SBR2 replies with the LOC binding update ACK to the MN (Step 12). In succession, the MN transmits a LOC Update message to CN, to notify its new LOC. CN updates the MN's LOC and responds with the LOC Update ACK message to the MN (Steps 13 and 14). CN can deliver the data packets to the MN via AR1, SBR1, SBR2 and AR4 in sequence.

We have looked over the existing mobility control schemes for ILNP-based mobile networks; i.e., ILNP-Global and ILNP-Local. We observe both the mobility schemes rely on the centralized DDNS server in some degrees. This results in poor scalability and performance in mobile environments.

Accordingly, we propose a new mobility control scheme in this paper, which manages the ID-LOC information in the fully distributed fashion.

3. Proposed Scheme

In this section, we describe our distributed mobility control scheme, named ILNP-Distributed, in detail.

3.1. Overview

We consider the mobility scenario of Figure 5 to discuss the operation of our proposed ILNP-Distributed scheme. In the scenario, MN (bob.SKT.com) and CN (alice.SKT.com) are in the same domain initially, and MN moves to another domain by roaming. Further, the MN changes the serving AR by handover in the visiting domain.

In the proposed scheme, each SBR has the dedicated mobile DDNS (m-DDNS) server and the server maintains two databases, called HHR and VHR. HHR keeps track of the ID-LOC (can be LLOC or GLOC) mappings for the home hosts, i.e., the hosts whose home domain is the SBR's serving domain, as shown in Table 1. VHR stores the ID-LLOC mappings for the hosts visiting the SBR's serving domain. In Table 2, we provide the example of VHR in LGU.com domain. We can use the host's IP address as the LLOC and the associated SBR's IP address as the GLOC. We manage the ID-LOC information similarly to the ILNP-Local, but we do not depend on the centralized DDNS server.

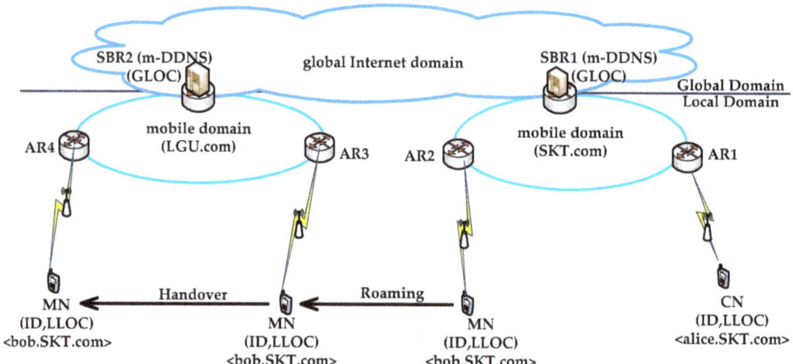

Figure 5. Mobility scenario for ILNP-Distributed scheme.

Table 1. Home Host Register (HHR) in SKT.com domain.

No.	Host Name	ID	LOC	Status
1	alice.SKT.com	ID1	LLOC	Home
2	bob.SKT.com	ID2	GLOC	Roaming
3	...	...	...	...

Table 2. Visiting Host Register (VHR) in LGU.com domain.

No.	Host Name	ID	LOC	Home Domain
1	bob.SKT.com	ID2	LLOC	GLOC
2	...	...	...	...
3	...	...	...	...

Before discussing the proposed scheme further, we compare our scheme with the existing mobility control schemes in the architectural perspective, as in Table 3.

Table 3. Comparison of candidate mobility control schemes.

Schemes	LOC	Mapping Architecture	Mobility Agent
ILNP-Global	GLOC	Centralized	DDNS
ILNP-Local	LLOC, GLOC	Centralized	SBR (LMT), DDNS
ILNP-Distributed	LLOC, GLOC	Distributed	SBR (m-DDNS)

The ILNP-Global employs only the GLOC as a LOC. A central DDNS server is used for the ID-GLOC mappings. The ID-GLOC binding and query operations are performed between each host and the DDNS server. The ILNP-Local is similar to the ILNP-Global. However, it has two types of LOCs, i.e., GLOC and LLOC, for each host. Each SBR performs the localized ID-LLOC mappings for the hosts in the serving domain, while it depends on the centralized DDNS server for the ID-GLOC information of the hosts in the visiting domains. In the ILNP-Distributed, we use LLOC and GLOC, as in ILNP-Local. Each SBR (or m-DDNS server) manages HHR and VHR databases. The ID-LOC information is managed by m-DDNS servers in the distributed manner without any centralized entities. We describe the detailed operations in the subsequent sections.

3.2. Procedures for Non-Roaming Scenario

When MN establishes the network connection with an AR in the home domain, it performs the ID-LOC binding and query operations as in Figure 6.

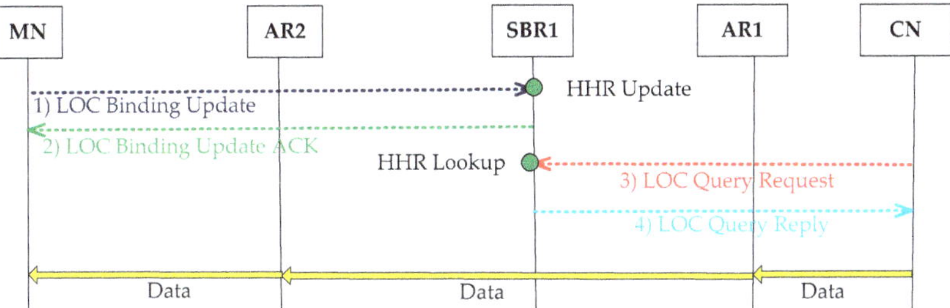

Figure 6. Binding and query operations for non-roaming.

MN registers the LLOC to SBR1 by sending a LOC binding update message (Step 1). The SBR1 stores the MN's ID-LLOC in the HHR, which maintains the ID-LOC (can be LLOC or GLOC) mappings of all the home hosts. Then the SBR1 returns the LOC binding update ACK message to the MN without further action (Step 2). When CN has data packets for the MN, CN queries the MN's LLOC to SBR1 by sending a LOC query request message (Step 3). Since the MN is in the home domain, SBR1 finds the LLOC in the HHR and returns it to the CN by the LOC query reply message (Step 4). CN then sends the data packets to the MN directly.

3.3. Procedures for Roaming Scenario

We illustrate the ID-LOC binding and query operations in Figure 7. When MN visits another domain (roaming case). When MN attaches to AR2 in the visiting domain, the MN sends a LOC binding update message to the SBR2 (Step 1), to register its LLOC. SBR2 then checks whether the MN belongs to its serving domain (home host) or not (roaming host). In the proposed scheme, we assume that each host has a name with its home domain information. For example, if a host has a name of alice.SKT.com, we can see its home domain is SKT.com, so the SBR2 recognizes that the MN is a roaming host, and stores the MN's ID-LLOC in VHR. In succession, SBR2 sends a LOC binding update message to the SBR1 in the MN's home domain (Step 2). SBR2 locates the SBR1 by using the m-DDNS server and the MN's home domain information. Then, SBR1 updates the MN's ID-GLOC mapping in the HHR and responds with the LOC binding update ACK to SBR2 (Step 3). The ACK message is eventually delivered to the roaming MN also by SBR2 (Step 4). When CN has data packets for the MN, the CN requests the MN's LOC to the SBR1 by sending a LOC query request message (Step 5). SBR1 finds the MN's GLOC in the HHR since the host is in the visiting domain, and returns it to the CN by the LOC query reply message (Step 6). With the GLOC, CN can send the data packets to the MN via AR1, SBR1, SBR2 and AR3 in sequence.

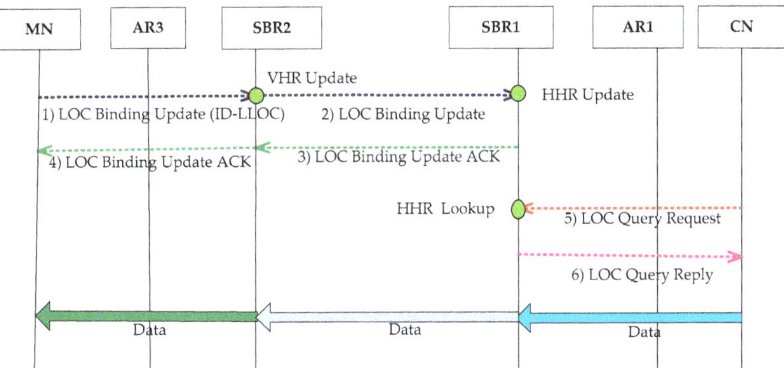

Figure 7. Binding and query operations for roaming.

3.4. Procedures for the Handover Scenario

We assume the proposed scheme exploits the link-layer information for handover, as defined in the IEEE 802.21 standard [18]. With the help of link-layer triggers, e.g. link-detected (LD) and link-up (LU), MN can tell if it has moved to a new network or not, while connected to the old network.

We show the ID-LOC management operation during the handover in Figure 8, under our ILNP-Distributed scheme. When MN detects a new network, it sends a LOC binding update message to SBR2 (Step 1), in order to register the LLOC in the VHR therein. After updating the ID-LLOC mapping, SBR2 replies to the MN with the LOC binding update ACK message (Step 2). We notice that the MN's GLOC does not change during handover. Hence the packets from CN arrive at SBR2 still. SBR2 looks up VHR and finds the LLOC of the MN. Then the packets are delivered to the MN via AR4 from SBR2.

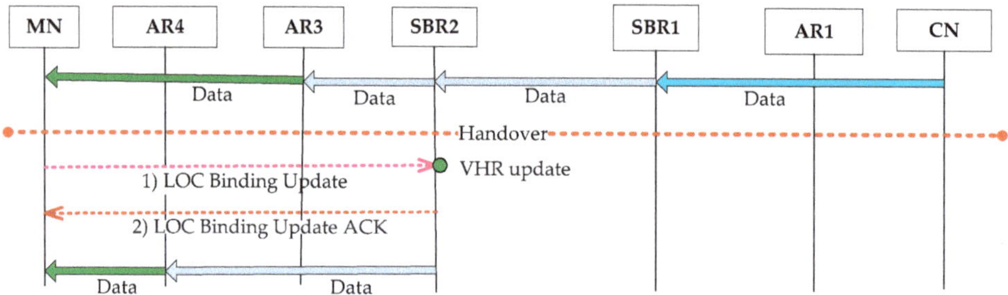

Figure 8. Handover operations.

4. Performance Analysis

In this section, we analyze the three candidate mobility control schemes for ILNP networks; i.e., ILNP-Global, ILNP-Local and INLP-Distributed. Especially, we evaluate their performances in terms of control traffic overhead, total transmission delay and handover delay.

4.1. Analysis Model

We consider the network model of Figure 9 to analyze the mobility control schemes. We first summarize the notations used for the analysis in Table 4 [15–17,19–21].

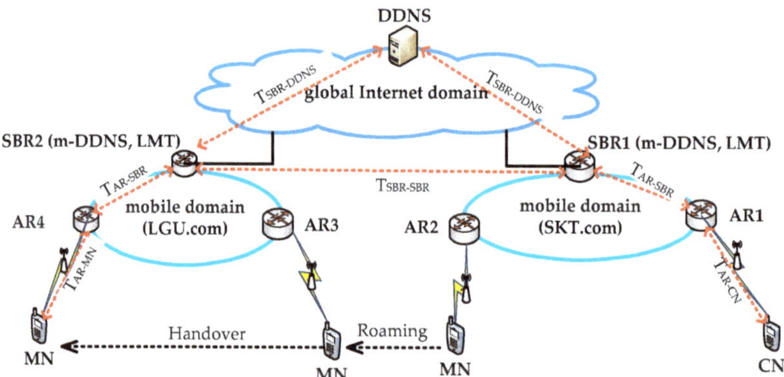

Figure 9. Network model for analysis.

Table 4. Parameters used for cost analysis.

Parameters	Description
S_c	Size of control packets (bytes)
S_d	Size of data packets (bytes)
B_w	Bandwidth of a wired link (Mbps)
B_{wl}	Bandwidth of a wireless link (Mbps)
L_w	Latency of a wired link (ms)
L_{wl}	Latency of a wireless link (ms)
$H_{x\text{-}y}$	Hop count between nodes x and y
α [17]	Unit cost for binding update
β [17]	Unit cost for database search
$N_{Host/AR}$	Number of hosts attached to an AR
N_{AR}	Number of ARs in the domain
N_{SBR}	Number of SBRs
q	Wireless link failure probability
Tq [15–17,19–21]	Average queuing delay at each router

We denote the transmission delay of a message with size S as $T_{X-y}(s)$ [15–17,19–21], where x and y are the sending and receiving nodes connected by a wireless link. We note that $T_{X-y}(s)$ can be written as

$$T_{X-y}(s) = \frac{1+q}{1-q} \times [\frac{S}{B_{wl}} + L_{wl}].$$

Similarly, we let $T_{X-y}(S, H_{X-y})$ [15–17,19–21] denote the transmission delay of a message with size S in the wired network, where x and y are the sending and receiving nodes connected by one or more wired links and H_{X-y} denotes the hop count between x and y. It is straightforward to write $T_{X-y}(S, H_{X-y})$ as

$$T_{X-y}(S, H_{X-y}) = \frac{1+q}{1-q} \times [\frac{S}{B_{wl}} + L_w + T_q].$$

4.2. Analysis of Control Traffic Overhead

We evaluate the control traffic overhead (CTO) for ID-LOC mapping managements at the SBRs and/or the DDNS server.

4.2.1. ILNP-Global

In the ILNP-Global scheme, we evaluate the CTO by the number of ID-LOC mapping messages the DDNS server handles. We consider the non-roaming case first. There, MN attaches to the serving AR, and registers its LLOC to the DDNS server by sending a LOC binding update message. Therefore, the DDNS server processes $S_C \times N_{Host/AR} \times N_{AR} \times N_{SBR}$ bytes of the control messages. Furthermore, each host sends a LOC query request message to the DDNS server for data delivery, which incurs additional $S_C \times N_{Host/AR} \times N_{AR} \times N_{SBR}$ bytes of control messages to the server. Therefore, the CTO for non-roaming case can be written as

$$
\begin{aligned}
CTO - NR_{ILNP-Global} \\
= S_C \times N_{Host/AR} \times N_{AR} \times N_{SBR} + S_C \times N_{Host/AR} \times N_{AR} \times N_{SBR}
\end{aligned}
\tag{1}
$$

In ILNP-Global, we note that the mobility control messages do not change for roaming and non-roaming cases. We, hence, obtain the CTO for roaming case as

$$CTO - RO_{ILNP-Global} = CTO - NR_{ILNP-Global} \tag{2}$$

4.2.2. ILNP-Local

In ILNP-Local, we first consider the non-roaming case, where we measure the CTO as the number of ID-LOC mapping messages processed by SBR1 and the DDNS server. When MN attaches to the AR in the home domain, the MN registers its LLOC to SBR1 by sending a LOC binding update message. That is, the SBR1 should handle $S_C \times N_{Host/AR} \times N_{AR}$ bytes of control messages. Subsequently, SBR1 sends a LOC binding update message to the DDNS server to register the MN's GLOC, so the DDNS server handles $S_C \times N_{Host/AR} \times N_{AR} \times N_{SBR}$ bytes of control messages from the SBRs. Accordingly, the CTO for non-roaming case is given as

$$CTO - NR_{ILNP-local} = S_C \times N_{Host/AR} \times N_{AR} + S_C \times N_{Host/AR} \times N_{AR} \times N_{SBR} \tag{3}$$

We now take into account the roaming case, where we evaluate the CTO by the amount of ID-LOC mapping messages at SBR2 and the DDNS server. When MN attaches to the AR in the visiting domain, the MN registers its LLOC to SBR2 by sending a LOC binding update message. Hence, the SBR2 handles $S_C \times N_{Host/AR} \times N_{AR}$ bytes of control messages. Then, the SBR2 sends a LOC binding update message to the DDNS server to update the MN's ID-GLOC information. For this, the DDNS server processes $S_C \times N_{Host/AR} \times N_{AR} \times N_{SBR}$ bytes of control messages. Further, when CN has packets for the MN, CN passes the packets to the SBR1 first. SBR1 sends the LOC query request message to the DDNS server to obtain the MN's GLOC, which incurs additional $S_C \times N_{Host/AR} \times N_{AR} \times N_{SBR}$ bytes of control messages. Thus, the CTO in roaming case can be written as

$$CTO - RO_{ILNP-local} = S_C \times N_{Host/AR} \times N_{AR} + S_C \times N_{Host/AR} \times N_{AR} \times N_{SBR} + S_C \times N_{Host/AR} \times N_{AR} \times N_{SBR} \tag{4}$$

4.2.3. ILNP-Distributed

For non-roaming case, we measure the CTO by the amount of ID-LOC mapping messages at SBR1. When MN attaches to the serving AR, it sends the LOC binding update messages to SBR1 and registers its LLOC. So SBR1 processes $S_C \times N_{Host/AR} \times N_{AR}$ bytes of control messages. Moreover, if CN has data packets for the MN, it sends a LOC query request message to the SBR1. Then the SBR1 should handle additional $S_C \times N_{Host/AR} \times N_{AR}$ bytes of control messages. Thus, we can write the CTO as

$$CTO - NR_{ILNP-Distributed} = S_C \times N_{Host/AR} \times N_{AR} + S_C \times N_{Host/AR} \times N_{AR}$$
$$CTO - NR_{ILNP-Distributed} = 2 \times S_C \times N_{Host/AR} \times N_{AR} \tag{5}$$

When MN visits a foreign domain (roaming case), we measure the CTO as the amount of ID-LOC mapping messages processed by SBR1 and SBR2. MN attaches to AR3 and registers its LLOC to SBR2 by sending a LOC binding update messages. Hence, the SBR2 processes $S_C \times N_{Host/AR} \times N_{AR}$ bytes of control messages. Successively, SBR2 sends a LOC binding update message to SBR1 to update the HHR entry. So the SBR1 handles $S_C \times N_{Host/AR} \times N_{AR}$ bytes of the control messages. Additionally, for data delivery, CN inquires the MN's LOC of SBR1 by sending a LOC query request message. Then, SBR1 processes $S_C \times N_{Host/AR} \times N_{AR}$ bytes of control messages. In total, the CTO for the roaming case is given as

$$CTO - RO_{ILNP-Distributed}$$
$$= S_C \times N_{Host/AR} \times N_{AR} + S_C \times N_{Host/AR} \times N_{AR}$$
$$+ S_C \times N_{Host/AR} \times N_{AR}$$
$$CTO - RO_{ILNP-Distributed} = 3 \times S_C \times N_{Host/AR} \times N_{AR} \tag{6}$$

4.3. Analysis of Total Transmission Delay

In this section, we derive the total transmission delay (TTD) for the ID-LOC management and data delivery. We notice that the TTD can be written as the sum of binding update delay (BUD), binding query delay (BQD) and data delivery delays (DDD), respectively. Hence, for each mobility scheme, we calculate the BUD, BQD and DDD separately and sum up them to obtain the TTD.

4.3.1. ILNP-Global

When MN attaches to the AR1 in the home domain (non-roaming case), it conducts the binding update with the DDNS server by exchanging the LOC binding update and ACK messages. This operation takes $2 \times (T_{MN\text{-}AR}(S_c) + T_{AR\text{-}SBR}(S_c, H_{AR\text{-}SBR}) + T_{SBR\text{-}DDNS}(S_c, H_{SBR\text{-}DDNS})) + P_{DDNS}$, where $P_{DDNS} = \alpha \log(N_{AR} \times N_{Host/AR} \times N_{SBR})$ represents the processing time in the DDNS server for the ID-GLOC update. We assume the processing time is proportional to the logarithm of the number of active hosts ($N_{AR} \times N_{Host/AR} \times N_{SBR}$), which can be attained with tree-based data structures. So, the BUD is written as

$$BUD - NR_{ILNP\text{-}Global}$$
$$= 2 \times (T_{MN\text{-}AR}(S_C) \times T_{AR\text{-}SBR}(S_C, H_{AR\text{-}SBR})$$
$$+ T_{SBR\text{-}DDNS}(S_C, H_{SBR\text{-}DDNS})) + P_{DDNS}$$

$$BUD - NR_{ILNP\text{-}Global} = 2 \times (T_{MN\text{-}AR}(S_C) \times T_{AR\text{-}SBR}(S_C, H_{AR\text{-}SBR}) +$$
$$T_{SBR\text{-}DDNS}(S_C, H_{SBR\text{-}DDNS})) + \alpha \log(N_{Host/AR} \times N_{AR} \times N_{SBR}).$$

In ILNP-Global, the binding update operation does not change regardless of the roaming and non-roaming cases. Hence, we obtain

$$BUD - RO_{ILNP\text{-}Global} = BUD - NR_{ILNP\text{-}Global}.$$

We next calculate the time for binding query operations. When CN has the data packets for MN in the home domain (non-roaming case), it sends a LOC query request message to the DDNS server to obtain the MN's GLOC. The DDNS server searches its database, which takes $P_{DDNS} = \beta \log(N_{AR} \times N_{Host/AR} \times N_{SBR})$, and responds to the CN by the LOC query reply message. This operation takes $2 \times (T_{CN\text{-}AR}(S_c) + T_{AR\text{-}SBR}(S_c, H_{AR\text{-}SBR}) + T_{SBR\text{-}DDNS}(S_c, H_{SBR\text{-}DDNS}))$. Thus, we have the BQD for non-roaming case as

$$BQD - NR_{ILNP\text{-}Global}$$
$$= 2 \times (T_{CN\text{-}AR}(S_C) \times T_{AR\text{-}SBR}(S_C, H_{AR\text{-}SBR})$$
$$+ T_{SBR\text{-}DDNS}(S_C, H_{SBR\text{-}DDNS})) + P_{DDNS}$$

$$BQD - NR_{ILNP\text{-}Global} = 2 \times (T_{CN\text{-}AR}(S_C) \times T_{AR\text{-}SBR}(S_C, H_{AR\text{-}SBR}) +$$
$$T_{SBR\text{-}DDNS}(S_C, H_{SBR\text{-}DDNS})) + \beta \log(N_{Host/AR} \times N_{AR} \times N_{SBR}).$$

When the MN is visiting another domain (roaming case), CN obtains the MN's GLOC from the DDNS server, following the same binding query operation. So, the BQD for roaming case is given as

$$BQD - RO_{ILNP\text{-}Global} = BQD - NR_{ILNP\text{-}Global}$$

We now investigate the data delivery of the ILNP-Global. If MN is in the home domain (non-roaming case), CN delivers data packets to the MN via AR1, SBR1 and AR2 in sequence. Hence the DDD is written as

$$DDD - NR_{ILNP\text{-}Global} = T_{CN\text{-}AR}(S_d) \times T_{AR\text{-}AR}(S_d, H_{AR\text{-}AR}) + T_{MN\text{-}AR}(S_d).$$

When MN is visiting another domain (roaming case), CN delivers data packets to the MN via AR1, SBR1, SBR2 and AR3 in sequence. So, the DDD can be written as

$$DDD - RO_{ILNP-Global} = T_{CN-AR}\left(S_d\right) \times T_{AR-SBR}\left(S_d, H_{AR-SBR}\right) + \\ T_{SBR-SBR}\left(S_d, S_d, H_{SBR-SBR}\right) + T_{AR-SBR}\left(S_d, H_{AR-SBR}\right).$$

In total, we derive the TTD for non-roaming case and the TTD for roaming case as

$$TTD - NR_{ILNP-Global} = BUD - NR_{ILNP-Global} + BQD - NR_{ILNP-Global} + DDD - NR_{ILNP-Global} \qquad (7)$$

$$TTD - RO_{ILNP-Global} = BUD - RO_{ILNP-Global} + BQD - RO_{ILNP-Global} + DDD - RO_{ILNP-Global} \qquad (8)$$

4.3.2. ILNP-Local

In ILNP-Local, the ID-LOC binding updates are performed as follows. When MN attaches to AR1 in the home domain (non-roaming case), it registers the LLOC to SBR1 by exchanging the LOC binding update and ACK messages. This operation takes $2 \times (T_{MN-AR}(S_c) + T_{AR-SBR}(S_c, H_{AR-SBR}) + P_{SBR}$, where $P_{SBR} = \alpha\log(N_{AR} \times N_{Host/AR})$ is the LMT update time at SBR1. Subsequently, SBR1 updates the MN's GLOC in the DDNS server by exchanging the LOC binding update and ACK messages. Considering the database update time PDDNS at the DDNS server, this operation takes $T_{SBR-DDNS}(S_c, H_{SBR-DDNS}))$ + P_{DDNS}, where $P_{DDNS} = \alpha\log(N_{AR} \times N_{Host/AR} \times N_{SBR})$. So, the BUD for non-roaming case can be written as

$$BUD - NR_{ILNP-Local}$$
$$= 2 \times \left(T_{MN-AR}\left(S_C\right) \times T_{AR-SBR}\left(S_C, H_{AR-SBR}\right)\right.$$
$$\left. + T_{SBR-DDNS}\left(S_C, H_{SBR-DDNS}\right)\right) + P_{SBR} + P_{DDNS}$$

$$BUD - NR_{ILNP-Local} = 2 \times \left(T_{MN-AR}\left(S_C\right) \times T_{AR-SBR}\left(S_C, H_{AR-SBR}\right) + \\ T_{SBR-DDNS}\left(S_C, H_{SBR-DDNS}\right)\right) + \alpha\log\left(N_{Host/AR} \times N_{AR} \times N_{SBR}\right) + \alpha\log\left(N_{Host/AR} \times N_{AR} \times N_{SBR}\right).$$

The binding update process is not affected by the roaming and non-roaming cases. Hence, the BUD for roaming case is given as

$$BUD - RO_{ILNP-Local} = BUD - NR_{ILNP-Local}$$

We then turn to the binding query delay. In ILNP-Local, the binding query is resolved in the SBR locally for the non-roaming case. Thus, the BQD is zero.

$$BQD - NR_{ILNP-Local} = 0.$$

When MN is in the foreign domain (roaming case), the binding query is processed as follows. First, CN passes the data packets to SBR1. SBR1 searches the LMT in vain and requests the MN's GLOC to the DDNS server by sending a LOC query request message. The DDNS server looks for the GLOC in its database, which takes $P_{DDNS} = \beta\log(N_{AR} \times N_{Host/AR} \times N_{SBR})$, and responds to SBR1 by the LOC query reply message. This operation takes $2 \times T_{SBR-DDNS}(S_c, H_{SBR-DDNS})$. So, the BQD for roaming case can be written as

$$BQD - RO_{ILNP-Local}$$
$$= 2 \times T_{SBR-DDNS}\left(S_C, H_{SBR-DDNS}\right)$$
$$+ \beta\log\left(N_{Host/AR} \times N_{AR} \times N_{SBR}\right)$$

Once the binding query is resolved, data packets are transmitted from CN to MN. When the MN is in the home domain (non-roaming case), data packets are delivered from CN to MN via AR1, SBR1 and AR2 in sequence, such that

$$DDD - NR_{ILNP-Local} = T_{CN-AR}\,(S_d)\,\times\,T_{AR-SBR}\,(S_d, H_{AR-SBR}) + T_{AR-SBR}\,(S_d, H_{AR-SBR}) + T_{MN-AR}\,(S_d).$$

If the MN is visiting another domain (roaming case), data packets from CN are delivered to MN via AR1, SBR1, SBR2 and AR3 in sequence. So the DDD is given as

$$DDD - RO_{ILNP-Local} = T_{CN-AR}\,(S_d)\,\times\,T_{AR-SBR}\,(S_d, H_{AR-SBR}) +$$
$$T_{SBR-SBR}\,(S_d, H_{SBR-SBR}) + T_{AR-SBR}\,(S_d, H_{AR-SBR}) + T_{MN-AR}\,(S_d).$$

Therefore, we obtain the TTD for non-roaming case and the TTD for roaming case as

$$TTD - NR_{ILNP-Local} = BUD - NR_{ILNP-Local} + BQD - NR_{ILNP-Local} + DDD - NR_{ILNP-Local} \qquad (9)$$

$$TTD - RO_{ILNP-Local} = BUD - RO_{ILNP-Local} + BQD - RO_{ILNP-Local} + DDD - RO_{ILNP-Local} \qquad (10)$$

4.3.3. ILNP-Distributed

In the ILNP-Distributed, when MN attaches to AR1 in the home domain (non-roaming case), it performs the ID-LOC binding operation with SBR1 by exchanging the LOC binding update and ACK messages. Then, SBR1 updates the HHR entry. This operation takes $2 \times (T_{MN-AR}(S_c) + T_{AR-SBR}(S_c, H_{AR-SBR})) + P_{SBR}$, where $P_{SBR} = \alpha log(N_{AR} \times N_{Host/AR})$. Hence, the BUD for non-roaming case can be written as

$$BUD - NR_{ILNP-Distributed} = 2 \times (T_{MN-AR}\,(S_C)\,\times\,T_{AR-SBR}\,(S_C, H_{AR-SBR})) +$$
$$P_{SBR} BUD - NR_{ILNP-Distributed} = 2 \times (T_{MN-AR}\,(S_C)\,\times\,T_{AR-SBR}\,(S_C, H_{AR-SBR})) +$$
$$\alpha \log\,(\,N_{Host/AR}\,\times\,N_{AR}\,\times\,N_{SBR}).$$

When MN is not in the home domain (roaming case), the MN conducts the binding update operation with SBR2 by exchanging the LOC binding update and ACK messages. Incorporating the VHR update time at SBR2, this operation takes $2 \times (T_{MN-AR}(S_c) + T_{AR-SBR}(S_c, H_{AR-SBR}) + P_{SBR}$, where $P_{SBR} = \alpha log(N_{AR} \times N_{Host/AR})$. In succession, SBR2 performs the binding update operation with SBR1 by exchanging the LOC binding update and ACK messages. There, the SBR1 updates the ID-GLOC information in the HHR. This operation takes $T_{SBR-SBR}(S_c, H_{SBR-SBR})) + P_{SBR}$, where $P_{SBR} = \alpha log(N_{AR} \times N_{Host/AR})$. So, the BUD for roaming case is given as

$$BUD - RO_{ILNP-Distributed} = 2 \times (T_{MN-AR}\,(S_C)\,\times\,T_{AR-SBR}\,(S_C, H_{AR-SBR}) +$$
$$T_{SBR-SBR}\,(S_C, H_{SBR-SBR})) + P_{SBR} + P_{SBR}$$
$$BUD - RO_{ILNP-Distributed} = 2 \times (T_{MN-AR}\,(S_C)\,\times\,T_{AR-SBR}\,(S_C, H_{AR-SBR}) +$$
$$T_{SBR-SBR}\,(S_C, H_{SBR-SBR})) + \alpha \log\,(\,N_{Host/AR}\,\times\,N_{AR}\,\times\,N_{SBR}) + \alpha \log(\,N_{Host/AR}\,\times\,N_{AR} \times N_{SBR}).$$

We now calculate the BQD in the non-roaming case. First, CN sends a LOC query request message to SBR1 to find the MN's LLOC. SBR1 looks for the LLOC in the HHR, which takes $P_{SBR} = \beta log(N_{AR} \times N_{Host/AR})$, and responds to the CN by the LOC query reply. This takes $2 \times T_{CN-AR}(S_c) + 2 \times T_{AR-SBR}(S_c, H_{AR-SBR})$. Hence, we can write the BQD for non-roaming case as

$$BQD - NR_{ILNP-Distributed} = 2 \times T_{CN-AR}\,(S_C) + 2 \times T_{AR-SBR}\,(S_C, H_{AR-SBR}) + \beta \log\,(\,N_{Host/AR}\,\times\,N_{AR}).$$

In the ILNP-Distributed, the binding query process is the same for roaming and non-roaming cases. So, we have

$$BQD - RO_{ILNP-Distributed} = BQD - NR_{ILNP-Distributed}.$$

When MN is in the home domain (non-roaming case), CN can send data packets to the MN via AR1 and AR2 in sequence, once it acquires the MN's LOC. Hence, the DDD is given as

$$DDD - NR_{ILNP-Distributed} = T_{CN-AR}\ (S_d) \times T_{AR-AR}\ (S_d, H_{AR-AR}) + T_{MN-AR}\ (S_d).$$

If MN is in the foreign domain (roaming case), the data packets from CN are delivered to the MN via AR1, SBR1, SBR2 and AR3 in sequence, such that DDD is written as

$$DDD - RO_{ILNP-Distributed} = T_{CN-AR}\ (S_d) \times T_{AR-SBR}\ (S_d, H_{AR-SBR}) +$$
$$T_{SBR-SBR}\ (S_d, H_{SBR-SBR}) + T_{AR-SBR}\ (S_d, H_{AR-SBR}) + T_{MN-AR}\ (S_d).$$

Conclusively, we obtain the TTD for non-roaming case and the TTD for roaming case as

$$TTD - NR_{ILNP-Distributed} = BUD - NR_{ILNP-Distributed} + BQD-$$
$$NR_{ILNP-Distributed} + DDD - NR_{ILNP-Distributed} \tag{11}$$

$$TTD - RO_{ILNP-Distributed} = BUD - RO_{ILNP-Distributed} + BQD-$$
$$RO_{ILNP-Distributed} + DDD - RO_{ILNP-Distributed} \tag{12}$$

4.4. Analysis of Handover Delay

We here analyze the handover delay (HOD) of the three candidate mobility control schemes.

4.4.1. ILNP-Global

Completing the handover to AR4, MN updates its GLOC in the DDNS server by exchanges the LOC binding update and ACK messages. Considering the database update time at the DDNS server, it takes $2 \times (T_{MN-AR}(S_c) + T_{AR-SBR}(S_c, H_{AR-SBR}) + T_{SBR-DDNS}(S_c, H_{SBR-DDNS})) + P_{DDNS}$, where $P_{DDNS} = \alpha log(N_{AR} \times N_{Host/AR} \times N_{SBR})$. Then, the MN notifies the new GLOC to CN by exchanging the LOC Update and ACK messages, which takes the additional time of $2 \times (T_{MN-AR}(S_c) + T_{AR-SBR}(S_c, H_{AR-SBR}) + T_{SBR-SBR}(S_c, H_{SBR-SBR}) + T_{AR-SBR}(S_c, H_{AR-SBR}) + T_{CN-AR}(S_c))$. At last, CN delivers the data packets to the MN via AR1, SBR1, SBR2 and AR4, which takes $T_{CN-AR}(S_d) + T_{AR-SBR}(S_d, H_{AR-SBR}) + T_{SBR-SBR}(S_d, H_{SBR-SBR}) + T_{AR-SBR}(S_d, H_{AR-SBR}) + T_{AR-MN}(S_d)$.

Hence, in ILNP-Global, we obtain the HOD as

$$
\begin{aligned}
HOD_{ILNP-Global} = {} & 2 \times (T_{MN-AR}\ (S_C) + T_{AR-SBR}\ (S_C, H_{AR-SBR}) \\
& + T_{SBR-DDNS}\ (S_C, H_{SBR-DDNS})) + \alpha \log (N_{\frac{Host}{AR}} \times N_{AR} \times N_{SBR}) + 2 \\
& \times (T_{MN-AR}\ (S_C) + T_{AR-SBR}\ (S_C, H_{AR-SBR}) + T_{SBR-SBR}\ (S_C, H_{SBR-SBR}) \\
& + T_{AR-SBR}\ (S_C, H_{AR-SBR}) + (T_{CN-AR}\ (S_d) + T_{AR-SBR}\ (S_d, H_{AR-SBR}) \\
& + T_{SBR-SBR}\ (S_d, H_{SBR-SBR}) + T_{AR-SBR}\ (S_d, H_{AR-SBR}) + (T_{MN-AR}\ (S_d))
\end{aligned}
\tag{13}
$$

4.4.2. ILNP-Local

MN makes a handover to AR4 and updates its LLOC at SBR2, by exchanging the LOC binding update and ACK messages. With the LMT update time at SBR, i.e., PSBR, it takes $2 \times (T_{MN-AR}(S_c) + T_{AR-SBR}(S_c, H_{AR-SBR})) + P_{SBR}$, where $P_{SBR} = \alpha log(N_{AR} \times N_{Host/AR})$. The MN then notifies the LOC to CN by exchanging the LOC Update and ACK messages, which takes $2 \times (T_{MN-AR}(S_c) + T_{AR-SBR}(S_c, H_{AR-SBR}) + T_{SBR-SBR}(S_c, H_{SBR-SBR}) + T_{AR-SBR}(S_c, H_{AR-SBR}) + T_{CN-AR}(S_c))$. The CN delivers the data packets to the MN via AR1, SBR1, SBR2 and AR4 in sequence. The delivery time of each packet can be written as

$T_{\text{CN-AR}}(S_d) + T_{\text{AR-SBR}}(S_d, H_{\text{AR-SBR}}) + T_{\text{SBR-SBR}}(S_d, H_{\text{SBR-SBR}}) + T_{\text{AR-SBR}}(S_d, H_{\text{AR-SBR}}) + T_{\text{AR-MN}}(S_d)$, we obtain the HOD of ILNP-Local as

$$
\begin{aligned}
HOD_{ILNP-Local} = \quad & 2 \times \left(T_{MN-AR}\left(S_C\right) \times T_{AR-SBR}\left(S_C, H_{AR-SBR}\right) \right) \\
& + \alpha \log \left(N_{\frac{Host}{AR}} \times N_{AR} \times N_{SBR} \right) + 2 \times \left(T_{MN-AR}\left(S_C\right) \right. \\
& + T_{AR-SBR}\left(S_C, H_{AR-SBR}\right) + T_{SBR-SBR}\left(S_C, H_{SBR-SBR}\right) \\
& + T_{AR-SBR}\left(S_C, H_{AR-SBR}\right) + \left(T_{CN-AR}\left(S_d\right) + T_{AR-SBR}\left(S_d, H_{AR-SBR}\right) \right. \\
& + T_{SBR-SBR}\left(S_d, H_{SBR-SBR}\right) + T_{AR-SBR}\left(S_d, H_{AR-SBR}\right) + \left(T_{MN-AR}\left(S_d\right) \right)
\end{aligned}
\tag{14}
$$

4.4.3. ILNP-Distributed

In ILNP-Distributed, when MN changes the serving AR, it updates the LLOC in SBR2 by exchanging the LOC binding update and ACK messages. Letting PSBR denote the VHR update time, we express the operation time as $2 \times (T_{\text{MN-AR}}(S_c) + T_{\text{AR-SBR}}(S_c, H_{\text{AR-SBR}})) + P_{\text{SBR}}$, where $P_{\text{SBR}} = \alpha \log(N_{\text{AR}} \times N_{\text{Host/AR}})$. Then, SBR2 can deliver the data packets from CN to the MN with additional delay of $T_{\text{AR-SBR}}(S_d, H_{\text{AR-SBR}}) + T_{\text{AR-MN}}(S_d)$. We, hence, obtain the HOD of ILNP-Distributed as

$$
\begin{aligned}
HOD_{ILNP-Distributed} = \ & 2 \times \left(T_{MN-AR}\left(S_C\right) + T_{AR-SBR}\left(S_C, H_{AR-SBR}\right) \right) + \\
& \alpha \log \left(N_{\frac{Host}{AR}} \times N_{AR} \times N_{SBR} \right) + T_{AR-SBR}\left(S_d, H_{AR-SBR}\right) + \left(T_{MN-AR}\left(S_d\right) \right).
\end{aligned}
\tag{15}
$$

5. Numerical Results

We compare the performances of the considered mobility control schemes, i.e., ILNP-Global, ILNP-Local and ILNP-Distributed, based on the equations in the previous section. For numerical results, we use the parameter values in Table 5, as in [15–17,19–21].

Table 5. Default parameter values.

Parameter	Default	Minimum	Maximum
L_{wl} (ms)	10	1	55
q	0.1	0.1	0.8
T_q (ms) [15–17,19–21]	5	1	55
$H_{\text{AR-SBR}}$	10	1	55
$N_{\text{Host/AR}}$	500	100	10000
N_{AR}	30	1	45
$H_{\text{SBR-DDNS}}$	20	1	55
N_{SBR}		100	
$H_{\text{AR-AR}}$		$\sqrt{N_{\text{AR}}}$	
$H_{\text{SBR-SBR}}$		$\sqrt{N_{\text{SBR}}}$	
α [17]		0.1	
β [17]		0.2	
L_w		2 ms	
S_c		50 bytes	
S_d		200 bytes	
B_{wl}		11 Mbps	
B_w		100 Mbps	

5.1. Non-Roaming Case

We discuss the performance results for the non-roaming case.

5.1.1. Control Traffic Overhead (CTO)

We show the CTOs, i.e., amount of control messages processed at the SBRs and/or the DDNS server, of the considered mobility control schemes in Figures 10 and 11, for different $N_{\text{Host/AR}}$ and N_{AR} respectively. Our proposed scheme, ILNP-Distributed, has a significantly smaller CTO, compared to the existing mobility schemes. This is due to the fact that our scheme distributes the ID-LOC management messages over all the SBRs in the network. The performance gaps increase fast as the numbers of hosts and/or ARs increase. We also note that the ILNP-Local has a smaller CTO than the ILNP-Global because the ILNP-Local scheme does not conduct the binding query operations for the hosts in the home domain (non-roaming case).

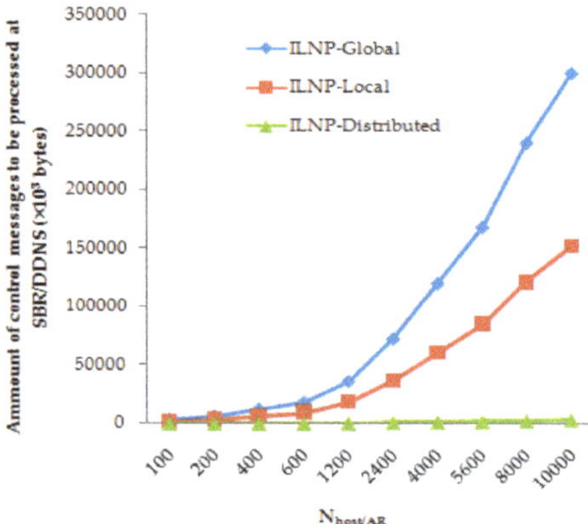

Figure 10. Impact of $N_{Host/AR}$ on CTO.

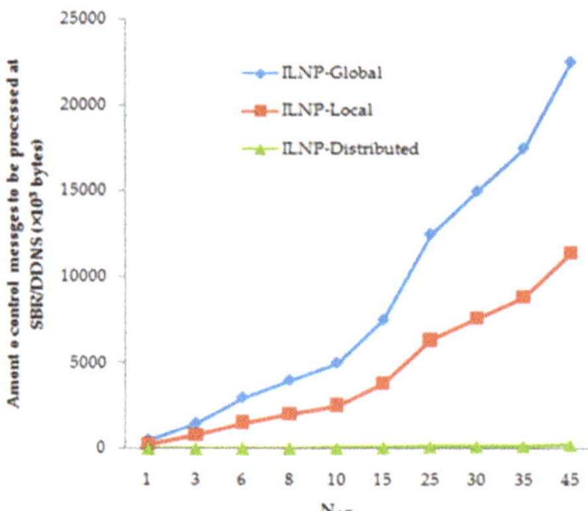

Figure 11. Impact of N_{AR} on CTO.

5.1.2. Total Transmission Delay (TTD)

In Figure 12, we demonstrate the impact of wireless link delay (L_{wl}) on the TTD. We observe that TTD increases steadily with L_{wl} for every mobility control scheme. However, among them, our proposed scheme shows the minimal delay. Further, we note that the ILNP-Local has smaller TTDs than the ILNP-Global since the ILNP-Local scheme does not perform the binding query operations for intra-domain data delivery.

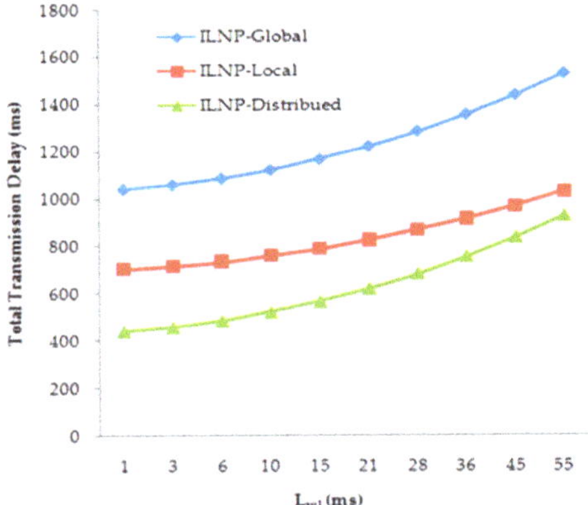

Figure 12. Impact of L_{wl} on TTD.

Next, we show the impact of wireless link failure probability (q) on the TTD in Figure 13. In all the mobility schemes, the TTD increases with q. The growth rate is low with small q but becomes high with large q. We can see that our ILNP-Distributed scheme has smaller TTDs for any fail probabilities, compared to the two existing schemes. Further, the ILNP-Local outperforms the ILNP-Global since the ILNP-Local scheme does not conduct the binding query operation for intra-domain data delivery.

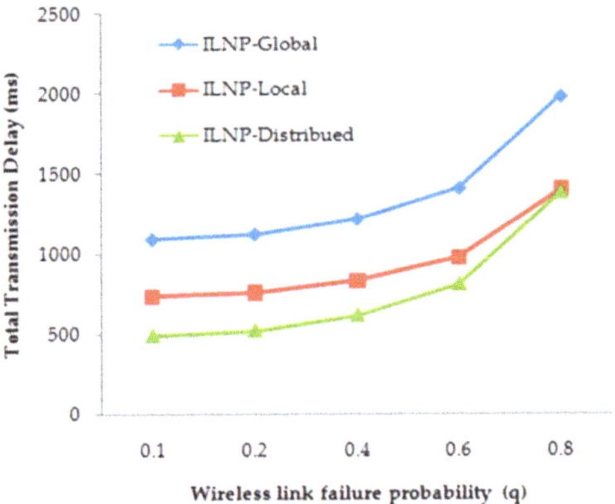

Figure 13. Impact of q on TTD.

In Figure 14, we observe the TTD curves while varying the average queuing delay (T_q) at each router. TTD increases quickly with larger T_q in all the candidate schemes. However, our proposed scheme shows the smallest value and the smallest growth rate among them. We note that the ILNP-Local is better than the ILNP-Global since the ILNP-Local scheme does not perform the binding query operation for intra-domain data delivery.

Figure 15 shows the impact of the hop count between AR and SBR ($H_{AR\text{-}SBR}$) on the TTD. TTD increases with the hop count for all the mobility schemes. Our ILNP-Distributed outperforms the competitors for any hop counts, and the ILNP-Local works better than the ILNP-Global.

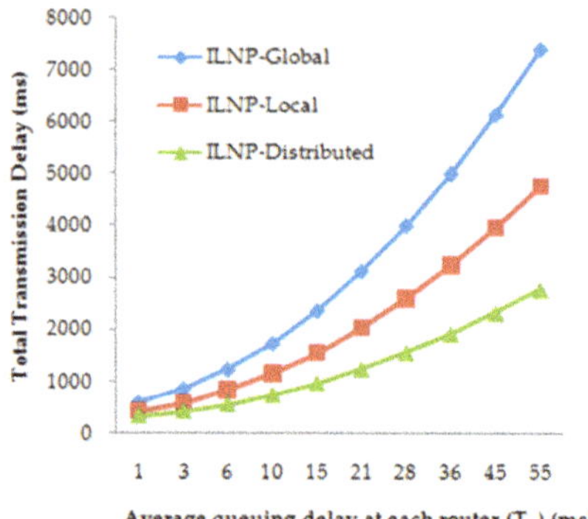

Figure 14. Impact of T_q on TTD.

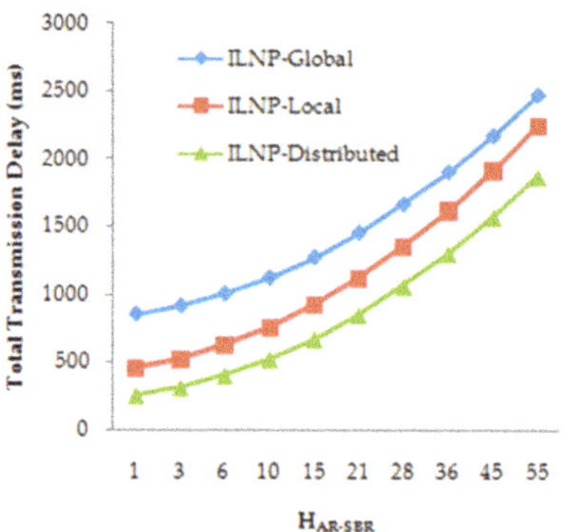

Figure 15. Impact of $H_{AR\text{-}SBR}$ on TTD.

We provide the TTD curves for different hop counts between SBR and the DDNS server ($H_{SBR\text{-}DDNS}$), in Figure 16. TTDs of the existing schemes are significantly affected by $H_{SBR\text{-}DDNS}$, since many binding messages are exchanged between SBRs and the DDNS server. On the contrary, our scheme does not employ the DDNS server and its TTD remains constant regardless of $H_{SBR\text{-}DDNS}$.

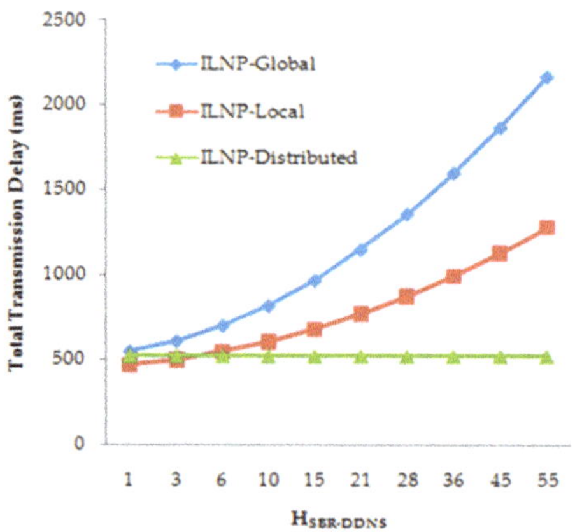

Figure 16. Impact of $H_{SBR\text{-}DDNS}$ on TTD.

5.2. Roaming Case

We here study the roaming case.

5.2.1. Control Traffic Overhead (CTO)

We present the CTOs for different $N_{Host/AR}$ and N_{AR}, in Figures 17 and 18 respectively. We observe that our proposed ILNP-Distributed shows a smaller CTO than the existing schemes, which can be explained as follows. In the previous mobility schemes, all the control messages are processed by the DDNS server. On the contrary, in our scheme, the control messages are distributed to the SBRs in the network. Their performance gaps increase rapidly with the number of hosts and/or ARs. We can see that the ILNP-Local performs poorly in roaming case and its CTO curves are overlapped with those of the ILNP-Global, due to the frequent binding update operations between each SBR and the DDNS server.

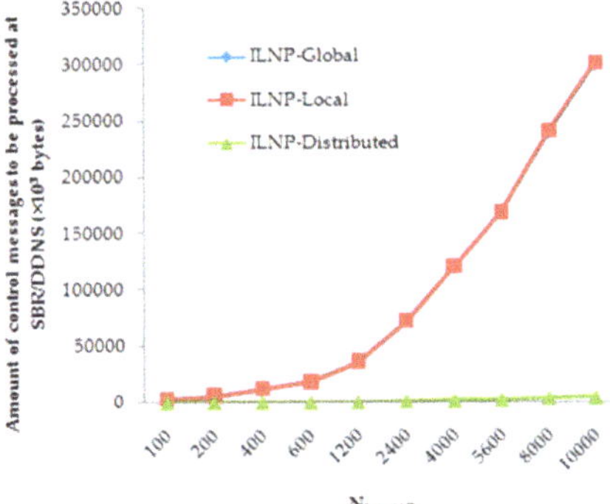

Figure 17. Impact of $N_{Host/AR}$ on CTO.

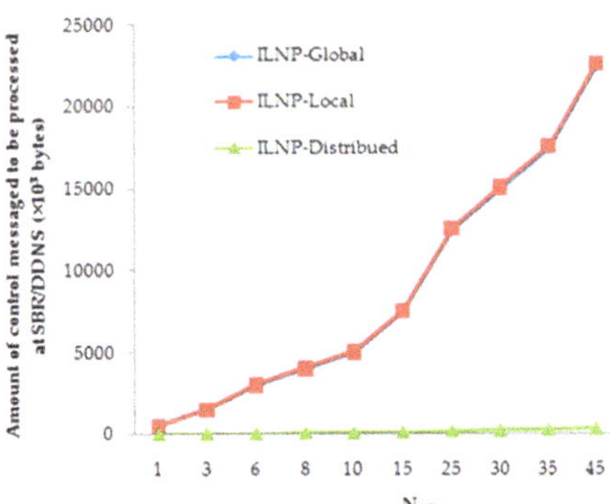

Figure 18. Impact of N_{AR} on CTO.

5.2.2. Total Transmission Delay (TTD)

We show the impact of wireless link delay (L_{wl}) on the TTD in Figure 19. The TTD increases steadily with L_{wl} in all the candidate mobility schemes. However, our ILNP-Distributed has, relatively, a smaller TTD than the existing schemes. Moreover, the ILNP-Local is better than the ILNP-Global. Their performance gap increases with L_{wl}.

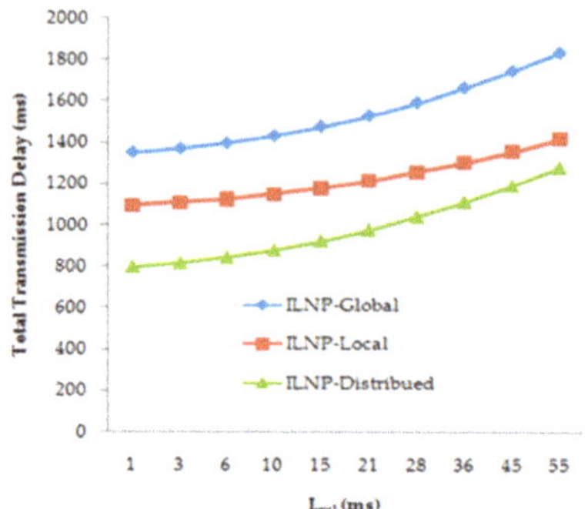

Figure 19. Impact of L_{wl} on TTD.

In Figure 20, we show the impact of wireless link failure probability (q) on the TTD. The TTD increases with q slowly for small value, but the growth rate increases for large failure probability. We can see that our ILNP-Distributed has the smallest TTD compared to the competitors. We also note that the ILNP-Local works better than the ILNP-Global since the ILNP-Local scheme does not perform the binding query operations for intra-domain data delivery.

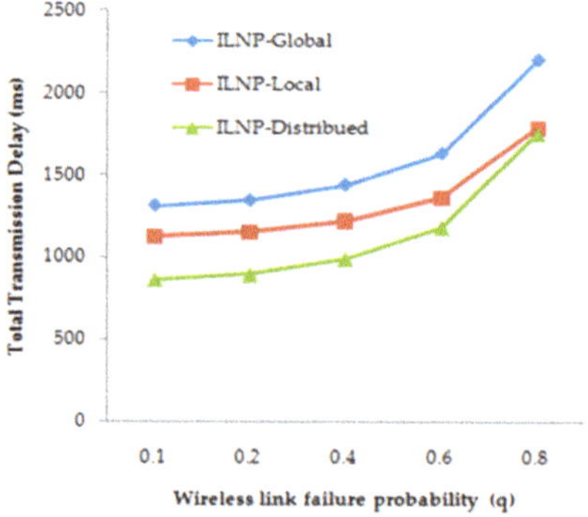

Figure 20. Impact of q on TTD.

In Figure 21, we compare the TTDs for average queuing delay (T_q) at each router. The TTD value increases with T_q for all the mobility control schemes, but our scheme has smaller values than the existing schemes for any T_q.

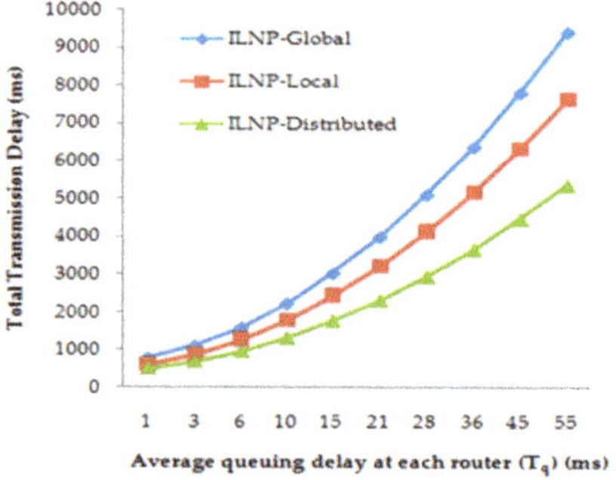

Figure 21. Impact of T_q on TTD.

We show the impact of the hop count between AR and SBR ($H_{AR\text{-}SBR}$) on the TTD, in Figure 22. We note that the TTD increases with the hop count for all the candidate schemes. Our proposed scheme works well when the hop count is small. However, as the hop count increases, it performs poorer than the ILNP-Local. This is because our ILNP-Distributed scheme conducts the binding query operations for data delivery between hosts and the SBR, for both roaming and non-roaming cases. In ILNP-Distributed the processing time have been done twice on PSBR.

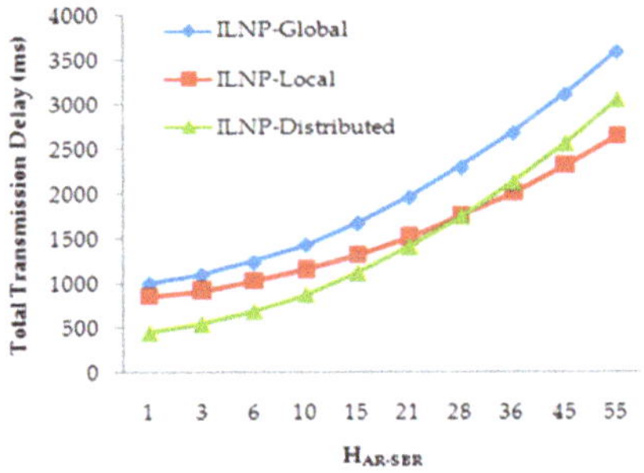

Figure 22. Impact of $H_{AR\text{-}SBR}$ on TTD.

Figure 23 compares the TTDs of the candidate schemes for different hop counts between SBRs and the DDNS server ($H_{SBR\text{-}DDNS}$). We can see that $H_{SBR\text{-}DDNS}$ has significant impacts on the TTD in the two existing schemes. In those schemes, the ID-LOC management messages are exchanged between SBRs and the DDNS server frequently. So, the two existing schemes depend on the centralized DDNS. The binding update and binding query operations perform with centralized DDNS.

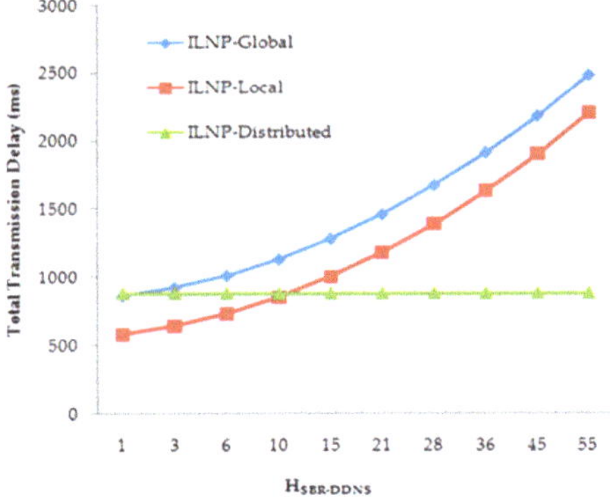

Figure 23. Impact of $H_{SBR\text{-}DDNS}$ on TTD.

5.3. Handover Delay (HOD)

In Figure 24, we present the impact of wireless link delay (L_{wl}) on the HOD. We observe the HOD increases with L_{wl} for all the candidate schemes. Our ILNP-Distributed scheme has a significantly smaller handover delay than the existing schemes. Moreover, the ILNP-Local works better than the ILNP-Global since the ILNP-Local conducts the binding update operations with the SBR only. We recall that in ILNP-Global, the binding updates after handover occurs between the hosts and the centralized DDNS server.

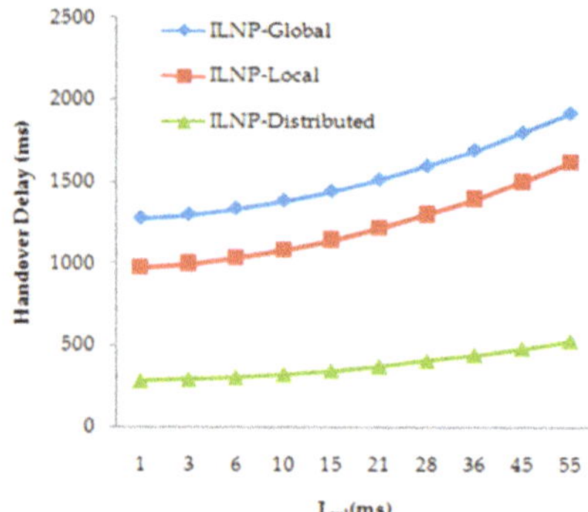

Figure 24. Impact of L_{wl} on HOD.

We show the impact of wireless link failure probability (q) on the HOD in Figure 25. For every mobility control scheme, the HOD increases with q, but our scheme performs much better than the existing schemes. The ILNP-Local gives smaller handover delay than the ILNP-Global due to the binding update operations within the domain.

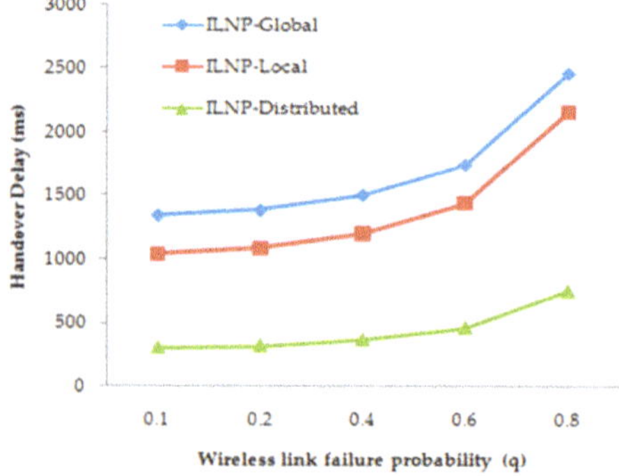

Figure 25. Impact of q on HOD.

We compare the HODs of candidate schemes while varying the average queuing delay (T_q) at each router, in Figure 26. The HOD increases with T_q for every mobility scheme. However, among them, our proposed scheme shows much better performance than any other schemes.

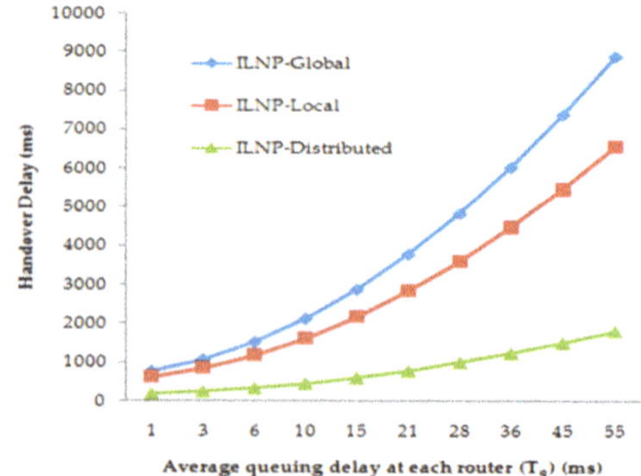

Figure 26. Impact of T_q on HOD.

In Figure 27, we demonstrate the impact of the hop count between AR and SBR ($H_{AR\text{-}SBR}$) on the HOD. We can see that the HOD increases with the hop count for all the mobility schemes, but the rate is relatively slow in our proposed scheme. The ILNP-Global and the ILNP-Local have a similar tendency while keeping a constant gap.

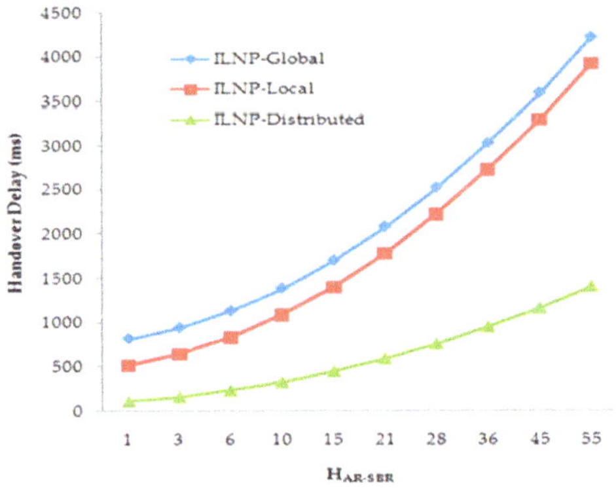

Figure 27. Impact of $H_{AR\text{-}SBR}$ on HOD.

5.4. Implementation Perspective

For applicability validation of the proposed scheme, we performed the testbed experimentation. For the testbed the locator was implemented by using the IPv4 address. The device ID was implemented by using the IPv6 address, which operates as 3.5 layer over IPv4.

The proposed scheme operations at GW can be summarized in Figure 28, which were implemented by using the Click Router software. When a packet arrives at the GW, it checks whether it is a control or data packet. If it is a control packet, the GW identifies the home GW for the device, based on the home domain information. Note that the home GW of device can be identified from the home domain information. In case that the packet is a data packet, if the GW already knows the LOC of the corresponding device, it will forward the packet directly. Otherwise, GW should first get the LOC of the device by contacting with the home GW of the device.

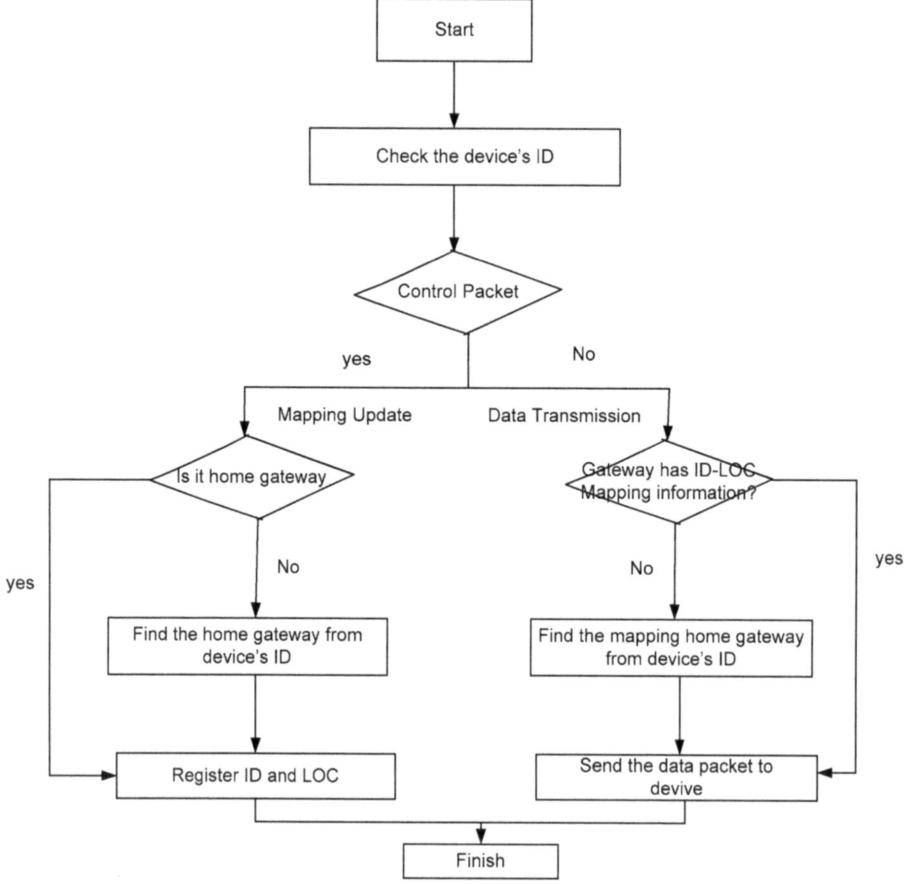

Figure 28. Algorithm at GW for mapping management and data transmission.

6. Conclusions

In this paper, we propose a new distributed mobility control scheme for mobile ILNP networks. Each SBR has the dedicated m-DDNS server, which maintains a Home Host Register (HHR) and a Visiting Host Register (VHR) to support the roaming. We let the m-DDNS trace all the ID-LOCs of the home hosts and the visiting hosts in the distributed manner. Hence, we obtain the LOC of any host of interest from the SBR (or m-DDNS server) of its home domain. By numerical analysis, we showed that our proposed scheme performs much better than the existing mobility control schemes in terms of control traffic overhead, total transmission delay and handover delay.

Author Contributions: M.G. wrote the initial manuscript; J.-G.C. and S.-J.K. revised the manuscript. M.M. and A.U.R. proofread the manuscript. W.A. conducted the performance analysis. All authors have read and agreed to the published version of the manuscript.

Funding: This research was supported in part by Basic Science Research Program through the National Research Foundation of Korea (NRF) funded by the Ministry of Education (2018R1D1A1B07048948), and in part by the 2019 Yeungnam University Research Grant.

Conflicts of Interest: The authors declare no conflict of interest.

List of Acronyms

Acronym	Description
ILNP	Identifier Locator Network Protocol
DDNS	Dynamic Domain Name Service
ID-LOCs	Identifier-locators
m-DDNS	mobile DDNS m-DDNS
HHR	Home Host Register
VHR	Visiting Host Register
IETF	Internet Engineering Task Force
SBR	Site Border Router
LLOC	Local Locator
GLOC	Global Locator
AR	Access Router
LMT	Local Mapping Table
LD	Link Detected
CTO	Control. Traffic Overhead
TTD	Total Transmission Delay
BUD	binding update Delay
BQD	Binding Query Delay
DDD	Data Delivery Delay
HOD	Handover Delay

References

1. Gohar, M.; Jung, H.; Koh, S.J. Distributed Mapping Management of Identifiers and Locators in Mobile-Oriented Internet Environment. *Int. J. Commun. Syst.* **2014**, *27*, 95–115. [CrossRef]
2. Morgan Stanley Report, Internet Trends. 2010. Available online: http://www.morganstanley.com/ (accessed on 28 September 2019).
3. Meyer, D.L.; Zhang, K. *Fall, Report from the IAB Workshop on Routing and Addressing*; IET: Fremont, CA, USA, 2007; RFC 4984.
4. Atkinson, R.J.; Bhatti, S.N.; Andrews, S.T. *Identifier-Locator Network Protocol (ILNP) Architecture Description*; IETF: Fremont, CA, USA, 2012; RFC 6740.
5. Atkinson, R.J.; Bhatti, S.N.; Andrews, S.T. *Optional Advances Deployment Scenarios for the Identifier-Locator Network Protocol (ILNP)*; IETF: Fremont, CA, USA, 2012; RFC 6748.
6. Atkinson, R.J.; Bhatti, S.N.; Andrews, S.T.; Rose, S. *DNS Resource Records for the Identifier-Locator Network Protocol (ILNP)*; IETF: Fremont, CA, USA, 2012; RFC 6742.
7. Atkinson, R.J.; Bhatti, S.N.; Stephen, H. ILNP: Mobility, multi-homing, localised addressing and security through naming. *Telecommun. Syst.* **2009**, *42*, 273–291. [CrossRef]
8. Atkinson, R.J. An Introduction to the Identifier-Locator Network Protocol (ILNP). In Proceedings of the IEEE London Communications Symposium (LCS), London, UK, 14–15 September 2006.
9. Atkinson, R.J.; Bhatti, S.N.; Stephen, H. Mobility as an Integrated Service Through the Use of Naming. In Proceedings of the 2nd ACM International Workshop on Mobility in the Evolving Internet (MobiArch), Kyoto, Japan, 27–30 August 2007.
10. Atkinson, R.J.; Bhatti, S.N.; Stephen, H. A Proposal for Unifying Mobility with Multi-Homing, NAT, & Security. In Proceedings of the 5th ACM International Workshop on Mobility Management and Wireless Access (MobiWAC), Chania, Crete, 22 October 2007.
11. Atkinson, R.J.; Bhatti, S.N.; Stephen, H. Mobility Through Naming: Impact on DNS. In Proceedings of the 3rd ACM International Workshop on Mobility in the Evolving Internet (MobiArch), Seatlle, WA, USA, 22 August 2008.
12. Atkinson, R.J.; Bhatti, S.N.; Stephen, H. Harmonized Resilience, Security and Mobility Capability for IP. In Proceedings of the 28th IEEE Military Communications Conference (MILCOM), Boston, MA, USA, 18–21 October 2009.

13. Rehunathan, D.; Atkinson, R.J.; Bhatti, S.N. Enabling Mobile Networks Through Secure Naming. In Proceedings of the 27th IEEE Military Communications Conference (MILCOM), San Diego, CA, USA, 16–19 November 2008.

14. Makaya, C.; Pierre, S. An analytical framework for performance evaluation of IPv6-based mobility management protocols. *IEEE Trans. Wirel. Commun.* **2008**, *7*, 972–983. [CrossRef]

15. Choi, N.J.; Gohar, M.; Koh, S.J. Domain-based distributed identifier-locator mapping management in Internet-of-Things networks. *Int. J. Netw. Manag.* **2018**, *28*, e2035. [CrossRef]

16. Kim, J.I.; Jung, H.; Koh, S.J. Mobile-Oriented Future Internet: Implementation and Experimentations over EU-Korea Testbed. *Electronics* **2019**, *8*, 338. [CrossRef]

17. Gohar, M.; Ahmed, W.; Bashir, F.; Choi, J.G.; Koh, S.J. A hash-based distributed mapping control scheme in mobile locator-identifier separation protocol networks. *Int. J. Netw. Manag.* **2017**, *27*, e1961. [CrossRef]

18. Atkinson, R.J.; Bhatti, S.N.; Stephen, H. Evolving the Internet Architecture Through Naming. *IEEE J. Sel. Areas Commun.* **2010**, *28*, 1319–1325. [CrossRef]

19. IEEE 802.21. *Local and Metropolitan Area Networks: Media Independent Handover (MIH)*; IEEE: Piscataway, NJ, USA, 2006.

20. Jung, H.; Gohar, M.; Kim, J.I.; Koh, S.J. Distributed mobility control in proxy mobile IPv6 networks. *IEICE Trans. Commun.* **2011**, *94*, 2216–2224. [CrossRef]

21. Gohar, M.; Koh, S.J. A distributed mobility control scheme in LISP network. *Wirel. Netw.* **2014**, *20*, 245–259. [CrossRef]

 *electronics*

Article

SAND/3: SDN-Assisted Novel QoE Control Method for Dynamic Adaptive Streaming over HTTP/3

Luis Guillen [1,*], Satoru Izumi [1], Toru Abe [1,2] and Takuo Suganuma [1,2]

[1]   Graduate School of Information Sciences, Tohoku University, Miyagi-Sendai 980-8577, Japan
[2]   Cyberscience Center, Tohoku University, Miyagi-Sendai 980-8577, Japan
*   Correspondence: lguillen@ci.cc.tohoku.ac.jp; Tel.: +81-22-217-5080

Received: 3 July 2019; Accepted: 1 August 2019; Published: 4 August 2019

Abstract: Dynamic Adaptive Streaming over HTTP (DASH) is a widely used standard for video content delivery. Video traffic, most of which is generated from mobile devices, is shortly to become the most significant part of Internet traffic. Current DASH solutions only consider either client- or server-side optimization, leaving other components in DASH (e.g., at the transport layer) to default solutions that cause a performance bottleneck. In that regard, although it is assumed that HTTP must be necessarily transported on top of TCP, with the latest introduction of HTTP/3, it is time to re-evaluate its effects on DASH. The most substantial change in HTTP/3 is having Quick UDP Internet Connections (QUIC) as its primary underlying transport protocol. However, little is still know about the effects on standard DASH client-based adaption algorithms when exposed to the future HTTP/3. In this paper, we present SAND/3, an SDN (Software Defined Networking)-based Quality of Experience (QoE) control method for DASH over HTTP/3. Since the official deployment of HTTP/3 has not been released yet, we used the current implementation of Google QUIC. Preliminary results show that, by applying SAND/3, which combines information from different layers orchestrated by SDN to select the best QoE, we can obtain steadier media throughput, reduce the number of quality shifts in at least 40%, increase the amount downloaded content at least 20%, and minimize video interruptions compared to the current implementations regardless of the client adaption algorithm.

Keywords: SDN; DASH; QoE; HTTP/3; QUIC

1. Introduction

Hypertext Transport Protocol (HTTP) version 3 or HTTP/3 is the name given by IETF to HTTP over Quick UDP Internet Connections (QUIC) [1]. HTTP/3 subsumes the benefits of HTTP/2 with TLS, reduced connection delay (0-RTT) on top of UDP, and at the same time offers TCP features (e.g., packet retransmission and congestion control). Being HTTP a fundamental component in Dynamic Adaptive Streaming over HTTP (DASH) [2], it is essential to analyze the effects on its overall performance for future internet traffic, which, according to the Visual Networking Index [3], is supposed to be mostly comprised by video streams from mobile devices.

It is worth noting that, although it is widely assumed that DASH must be transported on top of TCP, there are no specific restrictions on using other protocols. For example, Google reported a significant improvement in YouTube and Search services when deployed on top of QUIC [1]. However, little is still known about the impact of this protocol over other DASH schemes. While the authors of [4–7] showed a positive influence, others, such as Bhat et al. [8], conversely showed a detriment in the user perceived quality, also known as Quality of Experience (QoE).

The broad idea of DASH is that servers store different versions of video content, all of which are listed in an XML file called Media Presentation Description (MPD). Based on the network status and other parameters, clients can request a representation (version) of the video file. However, the DASH standard does not mandate a specific way to perform the adaption. Consequently, there are several approaches on how to adapt the video quality request, from hybrid quality adaption [9] to comprehensive optimization models [10]. Nonetheless, a single-parameter-based adaption (i.e., client or server) offers poor QoE. While a purely client-based adaption might not reach optimal performance due to its restricted knowledge, a purely server-side adaption would disregard users' features. Moreover, due to the ever-changing network conditions, frequent adaption requests based on inaccurate information lead to instability.

In that regard, using a centralized control plane would improve the video delivery [11,12]. Therefore, a Software Defined Networking (SDN)-based solution will adequately help to improve the adaption by providing a means of communication that includes an overview of the entire infrastructure. SDN allows the segregation of the control plane from the forwarding plane; therefore, due to its centralized nature and its global knowledge of the network, it can act as an intermediary between the DASH clients and servers, improving the adaption schemes. Therefore, the main goal of this paper is to evaluate the effects of SDN combined with HTTP/3 in DASH adaption schemes. To this aim, we implemented a holistic solution that considers all components involved in a DASH system to improve the overall QoE. The primary hypothesis to test was that, by applying the proposed approach of using HTTP/3 assisted by SDN and taking into account cross-layer collaboration, the overall QoE in DASH would significantly improve compared to current solutions that use single-component optimization on top of TCP.

The specific contribution of this paper is two-folded:

- In Section 3, we describe the overall design and interaction of the proposed QoE Control method for DASH over HTTP/3, whose main strength is its simplicity, and how SDN can orchestrate network-, transport-, and user-level components.
- In Section 5, we present a comprehensive evaluation of some of the most common adaption algorithms and the effects on QoE when they use HTTP/3 compared to the current deployment.

Based on preliminary results in Section 6, the proposed approach presents various advantages for DASH, i.e., higher and steadier average bitrate, minimizing the number of video stalls (freezes), and reducing the number of video quality shifts compared to traditional methods.

2. Related Work

2.1. TCP-Based SDN Solutions for DASH

Although there is a large body of knowledge about DASH and how to perform the adaption [11,13]. in this section, we briefly present the related work focused on SDN-based solutions to improve video delivery in DASH over traditional TCP. Some of the early work [14–16] proposed improving video quality by dynamically routing prioritized video stream flows over OpenFlow (OF) networks. Although using a dynamic priority route segregates the traffic, relying on network-only variables will be significantly constrained. In that regard, the authors of [17–19] improved the overall QoE by integrating *client feedback*, which constitutes an additional layer of information to perform the adaption. Therefore, by adding information from higher-layers into the adaption process will provide more flexible support to improve the user's QoE. The authors of [20,21] went even further and proposed an SDN-based architecture for DASH; both solutions presented innovative advantages for quality management. However, their deployment increased the network overhead. SDNHAS [22], an improved version of the method in [21], presents an elegant solution that combines multi-layer optimization based on user categories. However, they all are still constrained by simply relying on HTTP over TCP at the transport layer.

2.2. Non-TCP-Based SDN Solutions for DASH

Relying HTTP upon single TCP connections brings along known problems (e.g., increased delay and head-of-line blocking). Therefore, methods that use variations at the transport level show an improvement in the overall QoE in DASH. For instance, the authors of [23,24] used Multipath TCP (MPTCP) [25], an extension of TCP, to transmit *subflows* of DASH streams over multiple paths. The results were significantly better than using single TCP connections. However, despite the benefits of using MPTCP, both end-points need kernel modification. Furthermore, there are other conditions to fulfill, which will over-complicate the deployment in DASH or even have no effect, as reported by James et al. [26]. Few other authors experimented with QUIC and DASH assisted by SDN. Hussein et al. [7] proposed a QUIC-aware SDN architecture to handle congestion control, load balancing, and security limitations of QUIC. Although it was one of the first to test the integration of these technologies, it is not clear how the proposed architecture help DASH systems, apart from the initial connection. Hayes et al. [5] also marginally used SDN with MPTCP and QUIC to reduce the initial connection and an adaptive protocol selection based on performance feedback. Although it is an interesting scheme, the constant exchangeability between protocols would create a service disruption in highly active services, which will lead to detrimental QoE. As observed, related work present innovative and practical solutions to improve video QoE in DASH using SDN; however, there are still some open issues summarized as follows:

- *(P1) Single-component optimization*: Most solutions focus on a single component optimization (network, client, or server).
- *(P2) Transport ossification*: Most solutions assume TCP as the only protocol for end-to-end communication for HTTP, which creates a performance bottleneck.

Our work complements and extends the existing work by holistically tackling these issues. Specifically, as opposed to the methods in Section 2.1, by using HTTP/3 as the transport protocol, the proposed method does not suffer from the inherent problems of using HTTP over TCP, i.e., the head-of-line (HOL) blocking issue. Moreover, complementing the efforts of the methods in Section 2.2, our proposed method includes user-level information to perform a more fine-tuned adaption.

3. SDN-Assisted Novel DASH QoE Control Method over HTTP/3 (SAND/3)

To solve the problems stated above and offer users a better QoE, in this section, we describe our SDN-Assisted Novel QoE control method for DASH over HTTP/3 (SAND/3). The overall architecture of the proposed scheme is depicted in Figure 1. As observed, SAND/3 combines user-, device-, service- and network-level information, handled by three modules whose function is as follows:

- *User module*: This module collects the end-user identity and the associated devices from a third-party service in which the user is subscribed (e.g., Netflix) and stores them in a repository called *user profile*. We assume that this information will be available and provided by the third-party service. The profile consists of a registry of tuples comprised of a user identification number, associated devices, and preferences. Thus, when a device requests a video segment throughout an HTTP message, the *device manager* fetches the associated profile and collects the basic specifications (e.g., display size, available memory, buffer size, type of device, and subscription plan) and assigns a user priority for the current video playback. For instance, a user with a premium subscription to Netflix streaming a video on a SmartTV would have a higher priority than a user with a free account on a mobile device since their quality requirements are different.
- *Network module*: This module performs traffic engineering, by routing the packets via the most suitable paths that offer the best Quality of Service (QoS). To do so, first, the network status is continuously monitored by the *Network Monitor* sub-module, which collects the statistics of all the elements in the network, allowing a better estimation of the available resources. As part of this module, the *Topology Manager* updates the network device status, and it is triggered every time there is a change in the topology. Finally, the *Routing Handler* calculates the end-to-end paths to handle the traffic according to the priority assigned to each user. For simplicity, we create *k*

different paths using a modified version of the well-known *k*-maximum disjoint paths Suurballe's algorithm [27], where *k* is the number of categories in the application policy, e.g., if the quality categories are high, normal, and low, then $k = 3$. The cost of the path (weight) is calculated using Dijkstra's shortest path algorithm, having each link a cost based on network parameters or a combination of them, i.e., available bandwidth, delay, packet loss and so forth. In the current implementation, we use the delay as the only factor to calculate the path cost, but these can be easily extended to more elaborate weights.

- *Application module*: Based on the user profile, the current state of the network, and the specific service policies, the QoE Manager sub-module recommends the most suitable settings for the transmission, which is handled by the Transport Handler sub-module. Note that the Transport Handler sub-module is in charge of performing the transport connection using QUIC protocol.

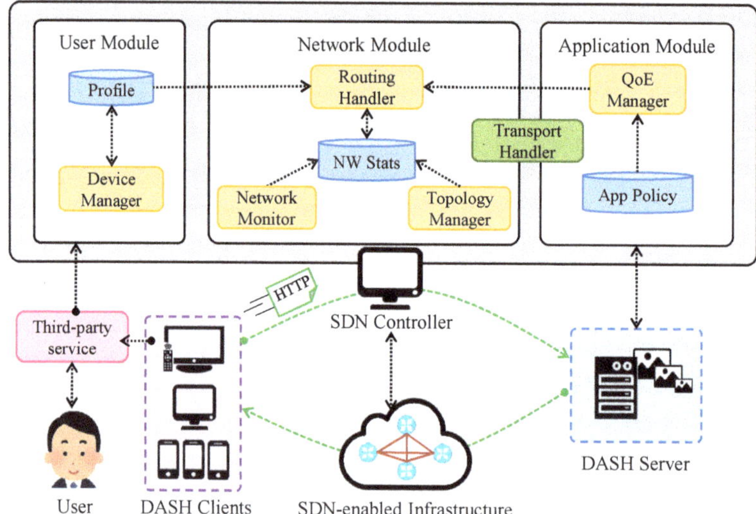

Figure 1. SAND/3 architecture.

4. An Overview of SAND/3

The network model is as follows, we consider a directed graph $G = (V, E)$, where V and E are the set of network devices and links connecting them, respectively. Each element in $v_i \in V$ is linked by an edge $e_{i,j} \in E$ between v_i and v_j. Moreover, $b_{i,j}$, $d_{i,j}$, and $c_{i,j}$ represent the bandwidth, delay, and cost, respectively of $e_{i,j}$. For simplicity, $c_{i,j}$ is calculated as the $b_{i,j} \times d_{i,j}$. If s and t are the source and destination devices, respectively, a path $p_{s,t} \in P$ is the shortest path from a source v_s to the destination v_t such that the cost $C_{s,t}$, the sum of link costs, is minimum. Finally, given a number of categories $k \in 1, 2, 3, ...$, the set P_k is comprised by the first k paths $p_{s,t}$.

The flowchart depicted in Figure 2 shows the overall process, which broadly works as follows: Firstly, when an HTTP request is received from the DASH client, it is captured at the edge switch and sent to the controller. Then, the system fetches the associated *User Profile* from the local repository (*Profile*) based on the information of HTTP headers—host, destination, user-agent, alt-svc, etc. In the current implementation, we assume single homogeneous devices for all users, and, therefore, we use the source and destination host as the identifier. However, more fine-grained information can be extracted from headers to match the request to a specific user properly. It is also important to note that, at this step, based on the information in the headers, the *Transport Handler* can determine whether both the client and server support HTTP/3 or use regular TCP connections. Then, the systems read the service policies from the local repository called *App Policy*, among those policies it is necessary to obtain the available user categories, which is then used for calculating paths with different priorities according to the user type. The mapping process from a user HTTP request to a path is shown in the dotted region in Figure 2. One of the most essential processes is the calculation of the user category, which can be done using various approaches, from advanced Machine Learning (ML) or other Artificial Intelligence (AI) as done by Bentaleb et al. [22], to other optimization heuristics as done by Herguner et al. [23]. However, for simplicity, in the current implementation, we attach this value as a parameter on the *User Profile*, leaving a more elaborate selection process as future work. Then, to create the paths, we calculate k (number of categories) paths so that the request is matched to the ith path, where i is the number of the category to which the user belongs. In case there is no available path, the traffic will be sent via the default shortest path. Finally, once the path has been selected and written in the adequate switches, the DASH client can start the transmission, and request/send them over the same connection until the playback is finished (see right-hand side of Figure 2). Note that the mapping is done once per transmission. However, it will be updated in the case there is a significant change in the topology (e.g., network device failure and link disconnection).

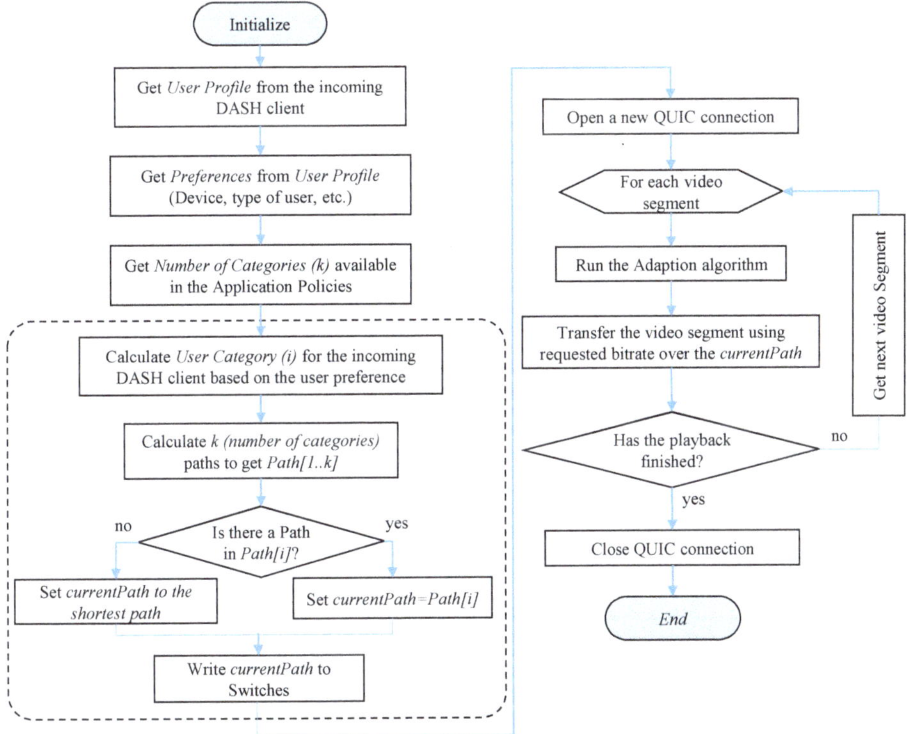

Figure 2. Overall flowchart process for QoE mapping to network segment from a source DASH client.

5. Evaluation

5.1. Test Environment

We used an emulated environment for both DASH and network, hosted on a single virtual machine running Ubuntu 16.04 LTS with 16 Gb of memory and six cores Intel Xeon(R) E5-2650 v4 of 2.20 GHz. The network was emulated using Mininet [28] v.2.2.2 which runs OpenFlow v.1.3 (OF) with the values described in Table 1. For simplicity, we used a 3×3 grid topology, as shown in

Figure 3. In the case of HTTP/3, since the official deployment was not available yet at the moment of the writing of this paper [29], we used the test QUIC client/server provided by Google [30]. The QUIC server runs the latest version (43 at the moment of the implementation), which, although it differs in some aspects to the IETF implementation, can serve as a preview on how the overall interaction of HTTP over QUIC would work on DASH. In the case of the DASH clients, we used a modified version of AStream [31], which is a Python-based emulated video player, with the QUIC client integrated as a sub-module as done by Arisu et al. [6]. Moreover, we modified the official QUIC server by automatically adding the required header tags (i.e., alternative protocol, port number, URL) per file in the server directory, which is mandatory in the test QUIC server. To compare with the current approach (using HTTP over TCP only), we used the built-in HTTP Server module from Python at the server side, and *libcurl* TCP client integrated into AStream. In addition, note that we kept the same connection for the entire transmission—as QUIC would do within the same connection—rather than opening a new connection per request. Finally, as the SDN controller, we used OpenDaylight (ODL) Beryllium SR2 and implemented the functions of each module as an internal application.

Table 1. Experimental parameters.

Parameter	Value
Topology	3×3 Grid
Bandwidth ($b_{i,j}$)	10 Mbps for all links
Delay ($d_{i,j}$)	Randomly assigned (1, 3, and 5 ms)
# of DASH Clients	3 (Xterm) emulating PCs
DASH Server	Simple HTTP Server
Controller	ODL Beryllium SR1

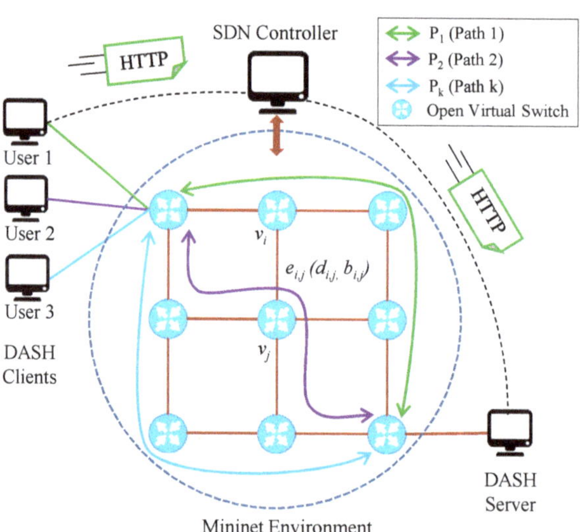

Figure 3. Testbed setup.

5.2. Use Case

Since streaming solutions over HTTP were originally developed for Video-on-Demand (VoD) [13], we used VoD as the use-case; however, we believe that the proposed solution can also be applicable for live streaming. To conduct the experiments, we used Big Buck Bunny (BBB) from a publicly available DASH dataset [32] by Lederer et al. [33]. In this dataset, the full-length BBB video (of approximately 600 s) is encoded in twenty different representations and segment (video chunks) sizes from 1 to

15 s ranging from bitrates of 420 × 360 pixels at 45 Kbps to 1920 × 1080 pixels at 3.9 Mbps. For the experiments, we used only a 4-s segment size, as it is an intermediate value between encoding and flexibility in the bitrate adaptation. To test the approach, we compared the performance of SAND/3 by using the following adaption bitrate (ABR) algorithms in the DASH clients:

- *Throughput-based adaption (TBA)* [34]: TBA starts the transmission by requesting the lowest available bitrate and, based on the average network throughput, the following segments will be selected in an additive increasing, multiplicative decreasing (AIMD) manner.

- *Buffer-Based adaption (BBA)* [35]: While other bitrate adaption approaches focus on throughput capacity prediction, BBA uses only the buffer occupancy to request the initial segment and then, if needed, estimates the capacity. BBA decreases the re-buffering events while the playback bitrate increases. The second version of this algorithm (BBA-2) was part of a large-scale experiment on Netflix, which achieved about 10–20% decreased re-buffering. Specifically for the experiments, after some preliminary results, instead of using a buffer size equal to the video segment size, as would usually be done, we used three times that value to make a fair comparison with the other compared algorithms. In Table 2, we describe a complete list of the BBA parameters used in the experiment.

- *Segment Aware Rate Adaptation (SARA)* [36]: Contrary to other ABR algorithms, which assume equal segment sizes, SARA takes into account the variation in segment and buffer sizes, so that it allows a more accurate prediction of the next segment. The configuration parameters used in the experiment are described in Table 3.

Table 2. List of BBA parameters used in the experiment.

Parameter	Value
Buffer size (B_{max})	12
Initial buffer	2
Initial factor	0.75
Reservoir (r)	0.2
Cushion (cu)	0.75

Table 3. List of SARA parameters used in the experiment.

Parameter	Value
Sample count	5
Initial buffering occupancy (B_{curr})	5
Re-buffering bound I	1
Buffer lower threshold (B_α)	5
Buffer upper threshold (B_β)	10

Each client was assigned a profile and a single device with the same specifications regarding the memory and CPU, emulated by Xterm terminals in Mininet. The priority level was different for each user; for simplicity, we numbered the priorities in ascending order so that high-, medium-, and low-priority would map to 0, 1, and 2, respectively. Note that, in the current implementation, these values were set statically in the user profile; a more comprehensive solution would require, for instance, a classification based on all the parameters, as done by Bentaleb et al. [22]. The experiment was conducted as follows. For each client ABR algorithm (namely TBA, SARA, and BBA), we played BBB on all the devices randomly starting within an approximate 30 s interval. In each run, all clients used the same adaption algorithm using the three variants:

- TCP-only approach: In this approach, the transmission of HTTP is over a TCP stream, as the current default solution would behave.

- QUIC-only approach: HTTP is sent on top of QUIC, emulating the way HTTP/3 would work but only changing the transport protocol without further improvements.

- SAND/3: The proposed scheme was used on top of QUIC and applying the process explained in previous sections.

In the first two cases, we left the default connectivity settings, which includes the end-to-end routing using simply the shortest path route. Note that, due to some performance limitations in the QUIC test server, not suitable for a large number of users [30], we only used three clients; nevertheless, since the primary objective of the experiment was to analyze the behavior and the feasibility of the proposal rather than a large-scale test, the current setup would suffice as a proof-of-concept.

5.3. QoE Metrics

To objectively measure the user QoE, there are five factors that influence the most in a DASH system: the number times the video freezes (stalls), the number of video quality shifts, media throughput (bitrate), start-up delay, and the fairness at shared bottlenecks [13,37]. However, since the start-up delay and fairness are mostly related to live-streaming, they are perceived less critical than the other variables [38]. Therefore, in this paper, we focus on the number of stalls, media throughput and the number of video quality shifts to measure the QoE.

6. Results

6.1. Number of Stalls

Concerning the number of stalls, which occurs when the video stops due to a buffer underrun, Table 4 shows the results obtained per adaption algorithm and their combined duration. As observed, there were stalls only in the TCP-only (current default) approach—the other approaches (QUIC and our approach) did not present any stalls during the playback whatsoever—which is a favorable result since a stall is the most detrimental factor for QoE [11]. In addition, note that most of the stalls occurred on users from the lower priority category. However, in the case of SARA, even users from the first categories suffered from stalls, and although there were few occurrences the combined time was considerable (16–26 s), which shows how unfair bandwidth competition can affect all users as the number of parallel requests increases.

Table 4. Number of stalls per user category using the TCP-only approach.

User Category	Adaption Algorithm					
	TBA		SARA		BBA	
	# of Stalls	Time [s]	# of Stalls	Time [s]	# of Stalls	Time [s]
1	0	0	2	21.1	0	0
2	0	0	1	16	0	0
3	1	20	4	26	1	2.46

6.2. Media Throughput

Concerning the media throughput (received video bitrate), depending on the adaption algorithm and the approach used, the results were relatively different. Figures 4–6 we show the media throughput obtained for each type of user using TBA, BBA, and SARA respectively.

Initially, in Figure 4a, we can see that, although TBA increased the requested bitrate progressively when using TCP-only approach, once higher quality segments were achieved and as new users started to request better quality representations, the video quality decreased significantly. This considerable quality decrease was due to the shared bandwidth bottleneck at the edge of the topology, and the unfair bandwidth competition. On the other hand, users that used QUIC-only, as shown in Figure 4b, received relatively low-quality segments for most of the transmission; however, at no time did the clients suffer from buffer underruns. Moreover, the bandwidth distribution was somehow evenly distributed among clients. Finally, in Figure 4c, for the case of our approach, the bitrate changes were quickly improved due to the characteristics of TBA. Moreover, once the clients converged to the best available video quality, the playback was steady and stayed, for the most part, in the highest video quality even for users with lower priority, which shows the benefits of the proposed approach for all user levels.

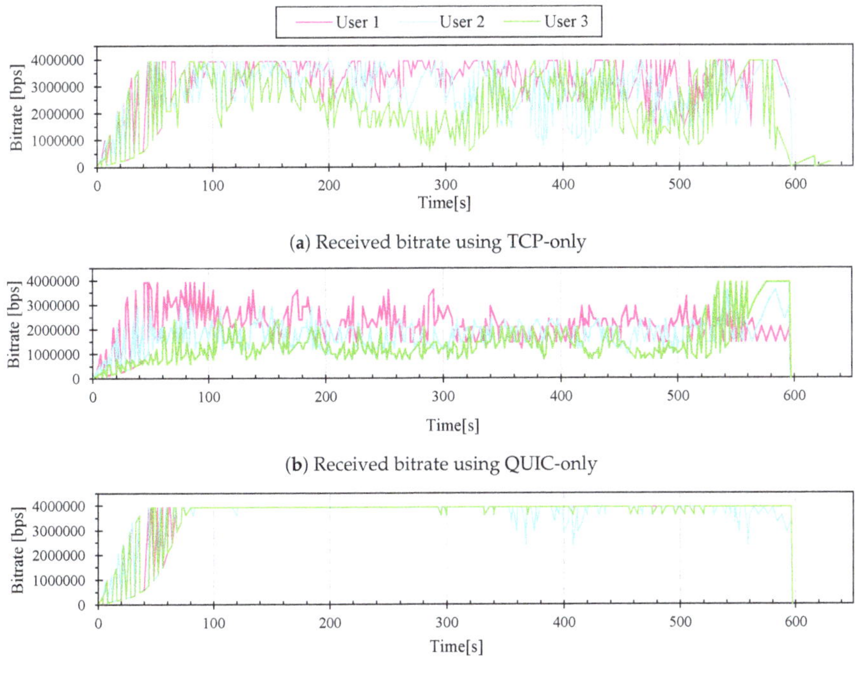

(**a**) Received bitrate using TCP-only

(**b**) Received bitrate using QUIC-only

(**c**) Received bitrate using the proposed approach

Figure 4. Media throughput obtained for three users at 10 Mpbs links using TBA adaption algorithm.

When SARA was used as the DASH adaption algorithm, as shown in Figure 5, since the algorithm considers the variable sizes of the different video representations, the convergence time was faster than the case of TBA. However, due to the congestion in the shared lines used in the TCP-, and QUIC-only approaches, the estimation was done using non-accurate information, and therefore the requested bitrate could not be handled by either of the approaches, leading to a highly unstable playback, as can be observed in Figure 5a,b. Due to the conservative nature of SARA, especially in terms of the *Delayed Download* mechanism, the video quality is gradually increased and maintained in the same one for a certain amount of time to avoid unnecessary downloads. However, by the time the segment is to be transmitted, the network conditions might have already changed so that the link will be overloaded downloading a bigger file until the threshold reaches the lower bound. On the other hand, by using our approach, once the optimal bitrate is selected, the feedback can estimate the available resources more accurately since the network status is continuously monitored on all network devices, which led to a steady and smooth playback with the highest video quality for all users, as shown in Figure 5c.

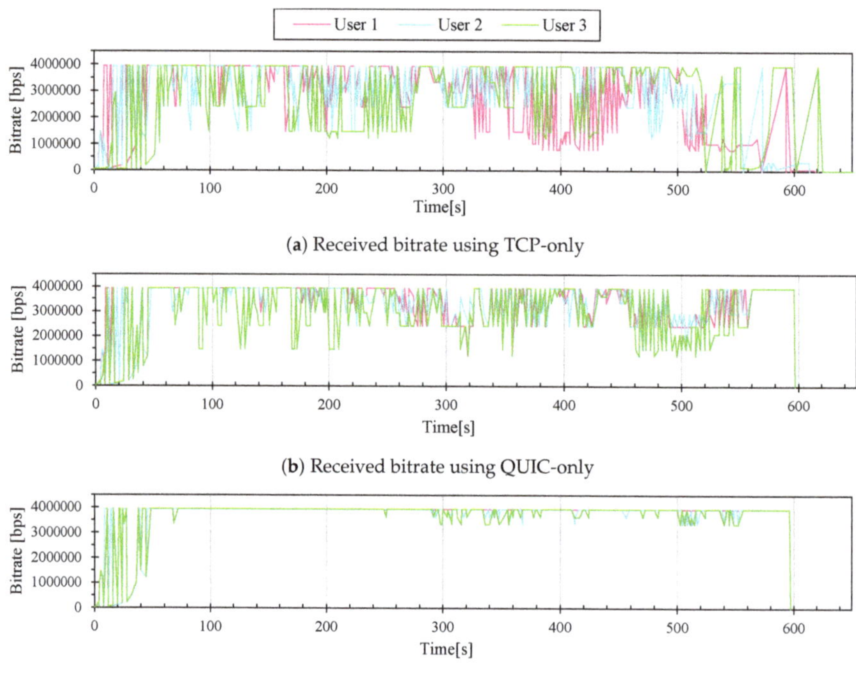

(**a**) Received bitrate using TCP-only

(**b**) Received bitrate using QUIC-only

(**c**) Received bitrate using the proposed approach

Figure 5. Media throughput obtained for three users at 10 Mpbs links using SARA adaption algorithm.

Finally, in Figure 6, we show the received bitrate when using BBA as the DASH adaption algorithm. Note that the bitrate variability is higher than the other algorithms. BBA has a different approach, which focuses on the current buffer occupancy rather than throughput estimation. Thus, it tends to request segments with higher bitrates for a longer time. However, when the streams compete with other requests, the congestion level increases considerably, leading to overloaded links and consequently higher loss-rate. As observed in Figure 6a, which depicts the received bitrate using TCP, there was a point in the playback (approximately at Second 250) where the congestion level was so high that the bitrate suffers from a phenomenon called *ON–OFF period*. In this state, the requested video quality oscillates between the highest and lowest, creating instability among the other DASH clients due to an inaccurate buffer estimation. In the case of QUIC, although there was some interference between Users 2 and 3, BBA was capable of recovering relatively quickly, as observed in Figure 6b. Lastly, in the case of our approach, as shown in Figure 6c, the bitrate was between the upper level of cushion (cu) and the maximum buffer capacity (B_{max}) for the most of the transmission, which allowed a smooth video playback without any buffer underruns.

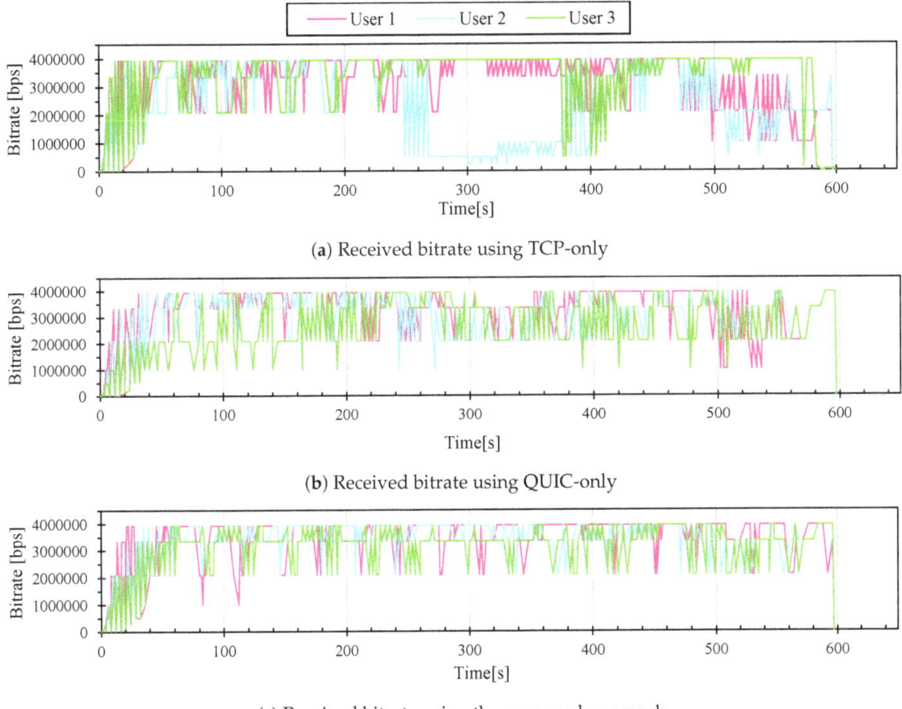

(**a**) Received bitrate using TCP-only

(**b**) Received bitrate using QUIC-only

(**c**) Received bitrate using the proposed approach

Figure 6. Media throughput obtained for three users at 10 Mpbs links using BBA adaption algorithm.

6.3. Video Quality Shifts

Concerning the number of video quality shifts, which refers to the change of video representation, users perceive dissatisfaction when the video abruptly changes to a considerably lower bitrate; therefore, a progressive increase of quality and minimizing the downshifts is desirable to have a good QoE. Figure 7 show the number of both up- and down-shifts events in every approach. As observed in Figure 7a–c, regardless on the adaption algorithm used (i.e., TBA, SARA, or BBA respectively), our approach was able to considerably minimize the downshift events at least for the user with the highest priority (User 1). However, note that, in the case of the number of the downshifts using BBA (Figure 7c), for Users 2 and 3, the number is higher using our approach than when using TCP. Nonetheless, as previously shown in Figure 6, the difference in the decrease of video quality was much bigger when using TCP than in our approach.

Regarding the number of upshifts, even though this variable is tightly coupled with the downshift events, the strategy used in each algorithm will significantly influence on how fast the bitrate will converge again to the best possible quality before a downshift event. Most of the adaption algorithms will opt for an aggressive decrease and progressive increase, having, therefore, more upshift events than downshifts, as shown in Figure 7d–f, which depict the number of upshift events using TBA, SARA, and BBA respectively. The average improvement achieved was around 45%, and 60% for the downshifts, and 40% and 50% for the upshift events compared to the TCP-only, and QUIC-only approaches, respectively, which is a significant improvement to the overall QoE.

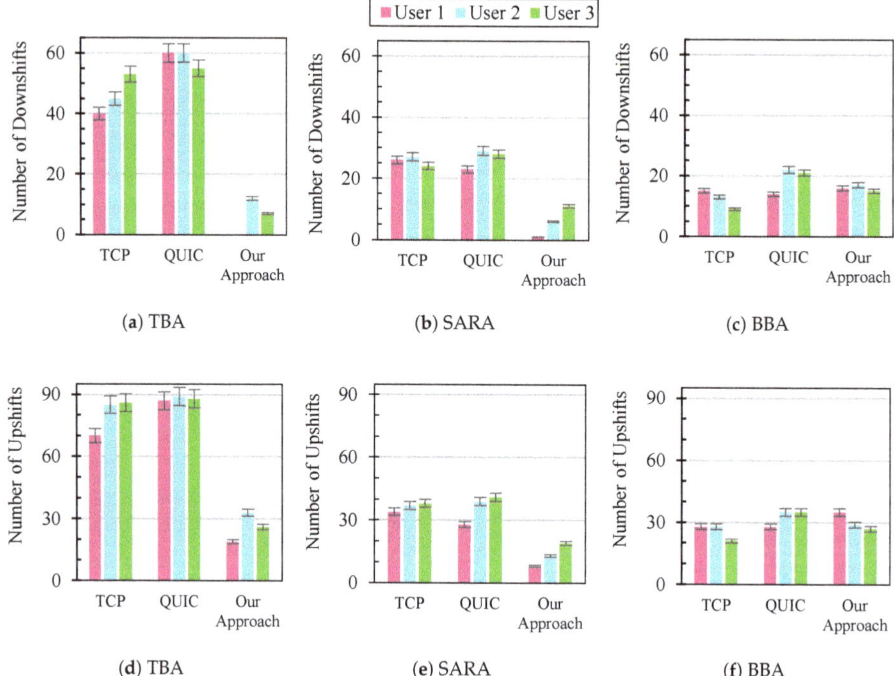

Figure 7. Number of up- and down-shifts per approach using different adaption algorithms.

6.4. Average Downloaded Video Files

Figure 8 depicts the heat-map of the average bitrate distribution of each of the approaches using the different adaption algorithms throughout the entire video. As observed, regardless of the algorithm, the bitrate distribution is considerably wider towards the highest video quality segments (light blue) when using the proposed approach. In fact, when using TBA and SARA, about 85% of the downloaded segments belonged to that category (Figure 8a,b), while, in the case of BBA (Figure 8c), it was about 45%, which was marginally better than the other approaches but still the second best video representation covered a larger part of the downloaded segments. This shows that it is possible to guarantee the best quality regardless of the user category or the adaption algorithm used. Since a better representation is stored in a bigger file, the amount of downloaded content was also significantly higher than when using TCP- or QUIC-only approaches, as shown in Figure 9a,b, respectively. For the most part, the distribution among the clients was rather even; however, note that there is a slight progressive decrease in the case of BBA (Figure 9c) as each user was affected by the number of clients and the increased congestion. Nevertheless, the approach achieved an average of 20% improvement compared to the TCP-only approach, and 25% compared to the QUIC-only approach.

These results confirm that the proposed approach provides a better overall QoE for all users in the different categories, while effectively using the available resources.

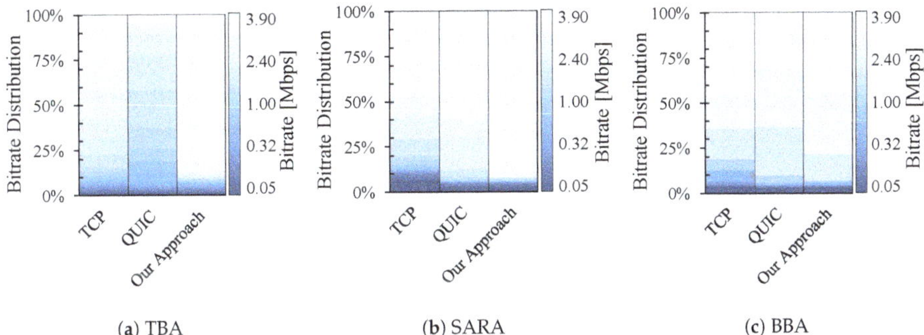

(a) TBA   (b) SARA   (c) BBA

Figure 8. Heat map of average downloaded video representation per approach.

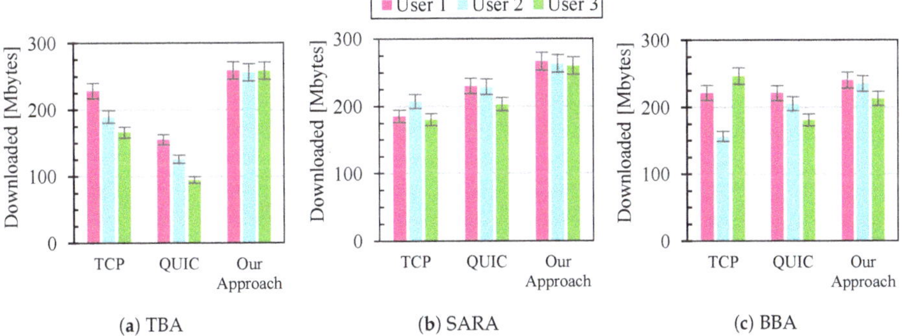

(a) TBA   (b) SARA   (c) BBA

Figure 9. Downloaded content per approach.

7. Discussion

Based on the obtained results, we could observe that using our approach, or even only implementing HTTP/3, will greatly benefit DASH performance. Even though in the current paper we used a test server from Google's QUIC implementation, the benefits can be extrapolated when the official IETF implementation becomes available. As observed, the main advantage would be the avoidance of stalling events regardless of the adaption algorithm used by the DASH client, even if they share a bottleneck at the edge of the topology or the network resources are scarce. This is, of course, an inherent benefit of using QUIC protocol at the transport level, which was precisely designed to improve the performance on users with constraint performance in terms of congestion, and loss percentage [1]. Changing just the transport protocol by itself will not be enough, as some other factors will also have a detrimental effect on the QoE, e.g., taking into account user categories or sudden changes in the network. However, although the inclusion of additional control levels will overload the network devices, there have been some recent proposals—e.g., *Segment Routing* (SR) [39]—that can help alleviate the flow-table exponential increase. The overall idea of SR is to divide the end-to-end path into multiple parts called "segments", which will be updated based on pre-established policies, and it is a useful technique to relief the flow-table overload in SDN implementations, as shown in [24].

An additional inherent advantage, especially for mobile users, would be solving the mobility issue. The control and identification of the transmission in QUIC is done by using a *Connection ID*, making therefore possible to migrate a transmission from an end-point that had a change in lower-layers (UDP, IP) and continue their process, even the in case of a vertical (i.e., interface) or horizontal (i.e., network) handover [29], as can be observed in [6] where the authors explored this phenomenon.

It is also important to mention that, although we did not modify the default values of TCP or UDP for the experiments (e.g., congestion window), we believe that, by adjusting this parameter, as suggested by Kakhki et al. [40], the results can be further improved for the approaches that use QUIC.

It is worth noting that using HTTP over QUIC has some drawbacks in the current implementations. For instance, Google and Facebook stated that a wide-scale deployment would require about twice the amount of CPU to cope with the traffic load compared to current HTTP/2 [29]. Of course, this performance issue might be reduced by further optimizing the hardware and software (i.e., when the changes are applied in the kernel) to the UDP stack in a rather short time, since this area is not as explored as in the case of TCP.

Another point to consider is that, since a centralized-control approach—which combines information otherwise segregated (i.e., user, network, service, and application)—might be seen challenging to implement, there are already authors who presented feasible solutions. For instance, Liotou et al. [41] proposed a solution that allows effective communication between Video Service Providers (VSPs) and mobile network operators' (MNOs), which enables feedback towards a better network-aware video segment selection.

Finally, note that we only focused on a fully SDN-based infrastructure; the current Internet is made of different intertwined technologies, and therefore the interaction between those non-SDN domains needs further study.

8. Conclusions and Future Work

In this paper, we present SAND/3, an SDN-Assisted QoE control method for DASH over HTTP3, which comprises a multi-layer collaborative optimization at the user, application, transport, and networking levels. The proposed approach combines the best state-of-the-art technologies to support quality adaption and improve DASH performance on top of HTTP/3. We implemented the proposed architecture over an SDN-based infrastructure using ODL as the controller in a Linux environment. Moreover, we evaluated its feasibility using an emulated DASH client using different adaption bitrate algorithms. Preliminary results show that, by considering end-user categories to manage the video segment's traffic over QUIC, the overall QoE improves not only regarding the media throughput but also reducing the number of stalls, and the number of abrupt downshifts of video quality compared to current TCP-only and QUIC-only approaches. However, although the QUIC-only approach showed no significant improvements for DASH by itself, we can conclude that QUIC (and HTTP/3) features might help to improve DASH performance but it needs further support.

Future directions this work might go are as follows: First, we still need to test the impact of the approach in different scenarios, e.g., using other adaption algorithms or heterogeneous clients. Then, we plan to deploy the system using larger network environments, and test using real implementations of DASH systems and HTTP/3. Finally, it might also be interesting to look at the impact of recent approaches, such as Multipath QUIC [42], as it could combine the benefits of QUIC streams multiplexing over multiple paths and their impact on future Internet applications, in particular video streaming.

Author Contributions: Conceptualization, T.S. and L.G.; investigation, data curation, methodology, software and writing—original draft preparation, L.G.; and writing—review and editing, and supervision, S.I., T.A. and T.S.

Conflicts of Interest: The authors declare no conflict of interest.

References

1. Langley, A.; Riddoch, A.; Wilk, A.; Vicente, A.; Krasic, C.; Zhang, D.; Yang, F.; Kouranov, F.; Swett, I.; Iyengar, J.; et al. The QUIC transport protocol: Design and internet-scale deployment. In Proceedings of the Conference of the ACM Special Interest Group on Data Communication (SIGCOMM'17), Los Angeles, CA, USA, 21–25 August 2017; pp. 183–196.

2. *ISO/IEC 23009-1:2014, Information Technology—Dynamic Adaptive Streaming over HTTP (DASH)—Part 1: Media Presentation Description and Segment Formats*; International Organization for Standardization: Geneva, Switzerland, 2014.

3. CISCO. *The Zettabyte Era: Trends and Analysis*; White Paper; CISCO, 2017. Available online: https://files.ifi.uzh.ch/hilty/t/Literature_by_RQs/RQ%20102/2015_Cisco_Zettabyte_Era.pdf (accessed on 3 July 2019)

4. Timmerer, C.; Bertoni, A. Advanced transport options for the dynamic adaptive streaming over HTTP. *arXiv* **2016**, arxiv:1606.00264.

5. Hayes, B.; Chang, Y.; Riley, G. Omnidirectional Adaptive Bitrate Media Delivery using MPTCP/QUIC over an SDN Architecture. In Proceedings of the IEEE Global Communications Conference (GLOBECOM2017), Singapore, 4–8 December 2017; pp. 1–6.

6. Arisu, S.; Begen, A.C. Quickly Starting Media Streams Using QUIC. In Proceedings of the 23rd Packet Video Workshop, Amsterdam, The Netherlands, 12–15 June 2018; pp. 1–6.

7. Hussein, A.; Kayssi, A.; Elhajj, I.H.; Chehab, A. SDN for QUIC: An Enhanced Architecture with Improved Connection Establishment. In Proceedings of 33rd Annual ACM Symposium on Applied Computing, Pau, France, 9–13 April 2018; pp. 2136–2139.

8. Bhat, D.; Rizk, A.; Zink, M. Not so QUIC: A Performance Study of DASH over QUIC. In Proceedings of the 27th Workshop on Network and Operating Systems Support for Digital Audio and Video, Taipei, Taiwan, 20–23 June 2017; pp. 13–18.

9. Mok, R.K.P.; Luo, X.; Chan, E.W.W.; Chan, R.K.C. QDASH: A QoE-aware DASH System. In Proceedings of the 3rd Multimedia Systems Conference, Chapel Hill, NC, USA, 22–24 February 2012; pp. 11–22.

10. Tashtarian, F.; Erfanian, A.; Varastehy, A. S2VC: An SDN-based Framework for Maximizing QoE in SVC-Based HTTP Adaptive Streaming. *Comput. Netw.* **2018**, *146*, 33–46. [CrossRef]

11. Seufert, M.; Egger, S.; Slanina, M.; Zinner, T.; Hoßfeld, T.; Tran-Gia, P. A survey on quality of experience of HTTP adaptive streaming. *IEEE Commun. Surv. Tutor.* **2015**, *17*, 469–492. [CrossRef]

12. Ganjam, A.; Jiang, J.; Liu, X.; Sekar, V.; Siddiqi, F.; Stoica, I.; Zhan, J.; Zhang, H. C3: Internet-scale control plane for video quality optimization. In Proceedings of the 12th USENIX Conference on Networked Systems Design and Implementation, Oakland, CA, USA, 4–6 May 2015; pp. 131–144.

13. Petrangeli, S.; van der Hooft, J.; Wauters, T.; De Turck, F. Quality of Experience-centric Management of Adaptive Video Streaming Services: Status and Challenges. *ACM Trans. Multimed. Comput. Commun. Appl.* **2018**, *14*, 31. [CrossRef]

14. Egilmez, H.E.; Civanlar, S.; Tekalp, A.M. An optimization framework for QoS-enabled adaptive video streaming over openflow networks. *IEEE Trans. Multimed.* **2013**, *15*, 710–715. [CrossRef]

15. Laga, S.; Van Cleemput, T.; Van Raemdonck, F.; Vanhoutte, F.; Bouten, N.; Claeys, M.; De Turck, F. Optimizing scalable video delivery through OpenFlow layer-based routing. In Proceedings of the 2014 Network Operations and Management Symposium (NOMS), Krakow, Poland, 4–9 May 2014; pp. 1–4.

16. Owens, H.; Durresi, A. Video over Software-Defined Networking (VSDN). In Proceedings of the 16th International Conference on Network-Based Information Systems, Gwangju, Korea, 4–6 September 2013; pp. 44–51.

17. Jarschel, M.; Wamser, F.; Hohn, T.; Zinner, T.; Tran-Gia, P. SDN-based Application-Aware Networking on the Example of YouTube Video Streaming. In Proceedings of the 2nd European Workshop on Software Defined Networks, Berlin, Germany, 4–6 October 2013; pp. 87–92.

18. Nam, H.; Kim, K.; Kim, J.; Schulzrinne, H. Towards QoE-aware Video Streaming using SDN. In Proceedings of the Communications Software, Services and Multimedia Symposium (Globecom2014), Austin, TX, USA, 8–12 December 2014; pp. 1317–1322.

19. Georgopoulos, P.; Elkhatib, Y.; Broadbent, M.; Mu, M.; Race, N. Towards Network-wide QoE Fairness Using OpenFlow-assisted Adaptive Video Streaming. In Proceedings of the Workshop on Future Human-Centric Multimedia Networking (FhMN13), Hong Kong, China, 16 August 2013; pp. 15–20.

20. Mkwawa, I.; Barakabitze, A.A.; Sun, L. Video Quality Management over the Software Defined Networking. In Proceedings of the IEEE International Symposium on Multimedia, San Jose, CA, USA, 11–13 December 2016; pp. 554–564.

21. Bentaleb, A.; Begen, A.C.; Zimmermann, R. SDNDASH: Improving QoE of HTTP Adaptive Streaming Using Software Defined Networking. In Proceedings of the 24th ACM International Conference on Multimedia, Amsterdam, The Netherlands, 15–19 October 2016; pp. 15–19.

22. Bentaleb, A.; Begen, A.C.; Zimmermann, R.; Harous, S. SDNHAS: An SDN-Enabled Architecture to Optimize QoE in HTTP Adaptive Streaming. *IEEE Trans. Multimed.* **2017**, *19*, 2136–2151. [CrossRef]

23. Herguner, K.; Kalan, R.S.; Cetinkaya, C.; Sayit, M. Towards QoS-aware Routing for DASH Utilizing MPTCP over SDN. In Proceedings of the 2017 IEEE Conference on Network Function Virtualization and Software Defined Networks (NFV-SDN), Berlin, Germany, 6–8 November 2017; pp. 1–7.

24. Barakabitze, A.A.; Mkwawa, I.; Sun, L.; Ifeachor, E. QualitySDN: Improving Video Quality using MPTCP and Segment Routing in SDN/NFV. In Proceedings of the International Conference on Network Softwarization (NetSoft2018), Montreal, QC, Canada, 25–29 June 2018; pp. 182–186.

25. Raiciu, C.; Paasch, C.; Barre, S.; Ford, A.; Honda, M.; Duchene, F.; Bonaventure, O.; Handley, M. How hard can it be? designing and implementing a deployable multipath TCP. In Proceedings of the 9th USENIX Symposium on Networked Systems Design and Implementation, San Jose, CA, USA, 25–27 April 2012; pp. 1–14.

26. James, C.; Halepovic, E.; Wang, M.; Jana, R.; Shankaranarayanan, N. K. Is Multipath TCP (MPTCP) Beneficial for Video Streaming over DASH? In Proceedings of the IEEE 24th International Symposium on Modeling, Analysis and Simulation of Computer and Telecommunication Systems (MASCOTS), London, UK, 19–21 September 2016; pp. 331–336.

27. Suurballe, J.W.; Tarjan, R.E. A quick method for finding shortest pairs of disjoint paths. *Networks* **1984**, *14*, 325–336. [CrossRef]

28. Lantz, B.; Heller, B.; McKeown, N. A Network in a Laptop: Rapid Prototyping for Software-Defined Networks. In Proceedings of the 9th ACM Workshop on Hot Topics in Networks, Monterey, CA, USA, 20–21 October 2010.

29. Stenberg, D. HTTP/3 Explained. Available online: Https://daniel.haxx.se/http3-explained/ (accessed on 3 July 2019).

30. Chromium-Playing with QUIC. Available online: Https://www.chromium.org/quic/playing-with-quic (accessed on 17 November 2018).

31. AStream: A Rate Adaptation Model for DASH. Available online: Https://github.com/pari685/AStream (accessed on 15 December 2018).

32. ITEC DASH Dataset. Available online: Http://www-itec.uni-klu.ac.at/ftp/datasets/DASHDataset2014/ (accessed on 23 January 2019).

33. Lederer, S.; Muller, C.; Timmerer, C. Dynamic adaptive streaming over HTTP dataset. In Proceedings of the 3rd ACM Multimedia Systems Conference, Chapel Hill, NC, USA, 22–24 February 2012; pp. 89–94.

34. Liu, C.; Bouazizi, I.; Gabbouj, M. Rate adaptation for adaptive HTTP streaming. In Proceedings of the 2nd ACM Conference on Multimedia systems, San Jose, CA, USA, 23–25 February 2011; pp. 169–174.

35. Huang, T.Y.; Johari, R.; McKeown, N.; Trunnell, M.; Watson, M. A buffer-based approach to rate adaptation: Evidence from a large video streaming service. In Proceedings of the ACM Conference (SIGCOMM'14), Chicago, IL, USA, 17–22 August 2014; pp. 187–198.

36. Juluri, P.; Tamarapalli, V.; Medhi, D. SARA: Segment aware rate adaptation algorithm for dynamic adaptive streaming over HTTP. In Proceedings of the Workshop on Quality of Experience-based Management for Future Internet Applications and Services, London, UK, 12–15 June 2015; pp. 1765–1770.

37. Timmerer, C.; Maiero, M.; Rainer, B.; Petscharnig, S. Quality of Experience of Adaptive HTTP Streaming in Real-World Environments. *IEEE COMSOC MMTC* **2015**, *15*, 1–4.

38. De Pessemier, T.; De Moor, K.; Joseph, W.; De Marez, L.; Martens, L. Quantifying the influence of rebuffering interruptions on the user's quality of experience during mobile video watching. *IEEE Trans. Broadcast.* **2013**, *59*, 47–61. [CrossRef]

39. IETF RFC-8402. Segment Routing Architecture. Available online: Https://tools.ietf.org/html/rfc8402 (accessed on 10 January 2019).

40. Kakhki, A.M.; Jero, S.; Choffnes, S.; Nita-Rotaru, C.; Mislove, A. Taking a Long Look at QUIC: An Approach for Rigorous Evaluation of Rapidly Evolving Transport Protocols. In Proceedings of the 2017 Internet Measurement Conference, London, UK, 1–3 November 2017; pp. 290–303.

41. Liotou, E.; Samdanis, K.; Pateromichelakis, E.; Passas, N.; Merakos, L. QoE-SDN APP: A Rate-guided QoE-aware SDN-APP for HTTP Adaptive Video Streaming. *IEEE J. Sel. Areas Commun.* **2018**, *36*, 598–615. [CrossRef]

42. De Coninck, Q.; Bonaventure, O. Multipath QUIC: Design and Evaluation. In Proceedings of the International Conference on Emerging Networking EXperiments and Technologies (Conext17), Incheon, Korea, 12–15 December 2017.

 *electronics*

Article

SmartX Box: Virtualized Hyper-Converged Resources for Building an Affordable Playground

Aris Cahyadi Risdianto, Muhammad Usman and JongWon Kim *

School of Electrical Engineering and Computer Science, Gwangju Institute of Science and Technology (GIST), Gwangju 61005, Korea; aris@nm.gist.ac.kr (A.C.R.); usman@smartx.kr (M.U.)
* Correspondence: jongwon@nm.gist.ac.kr; Tel.: +82-62-715-2219

Received: 19 September 2019; Accepted: 17 October 2019; Published: 30 October 2019

Abstract: In this paper, we present our proposals and efforts for building an affordable playground (i.e., miniaturized testbed) for Software-Defined Networking (SDN)-Cloud experiments by using hyper-converged SmartX Boxes that are distributed across multiple sites. Each SmartX Box consists of several virtualized functions that are categorized into SDN and cloud functions. Multiple SmartX Boxes are deployed and inter-connected through SDN to build multi-site distributed cloud playground resources. The resulting deployment integrates both cloud multi-tenancy and SDN-based slicing, which allow developers to run experiments and operators to monitor resources in a distributed SDN-cloud playground. It also describes how the hyper-converged SmartX Box can increase the affordability of the playground deployment. Thus, the analysis result shows the efficiency of SmartX Box for building a distributed playground by providing semi-automated DevOps-style resource provisioning.

Keywords: affordable playground; hyper-converged SmartX Box; distributed resources; multi-site and virtualized cloud; software-defined networking; DevOps automation

---

1. Introduction

1.1. Background

Aligned with worldwide Future Internet testbed efforts (e.g., GENI—Global Environment for Network Innovations [1], FIRE—Future Internet Research and Experimentation [2]), OF@TEIN (OpenFlow at Trans-Eurasia Information Network) project was started to build an OpenFlow-enabled testbed over TEIN infrastructure in 2012 [3]. Several experimentation tools were developed to support both developers and operators in using OF@TEIN testbed. Initially, a mixed combination of tools, ranging from simple web-/script-based to DevOps (Development and Operations) [4] Chef-based automated tools [5], were deployed over SmartX Racks. SmartX Rack consists of four devices: *Management & Worker node, Capsulator node, OpenFlow switch,* and *Remote power device.* Physically LAN-connected SmartX Racks were inter-connected by L2 (layer 2) tunnels, employed in Capsulator nodes as described in detail in this work [6]. However, SmartX Racks with multiple devices were subject to physical remote re-configurations, which are extremely hard to manage for distributed OF@TEIN Playground. Thus, from late 2013, a hyper-converged SmartX Box was introduced to virtualize and merge the functionalities of four devices into a single box [7]. The comparison between previously deployed SmartX Racks and newly deployed SmartX Box, is depicted in Figure 1. Finally, several hyper-converged SmartX Boxes were distributed deployed over nine Asian countries in 2015, as shown in Figure 2.

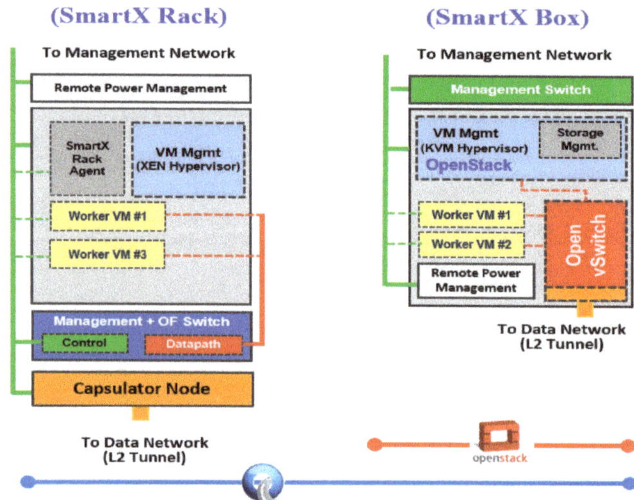

Figure 1. SmartX Rack versus SmartX Box Comparison.

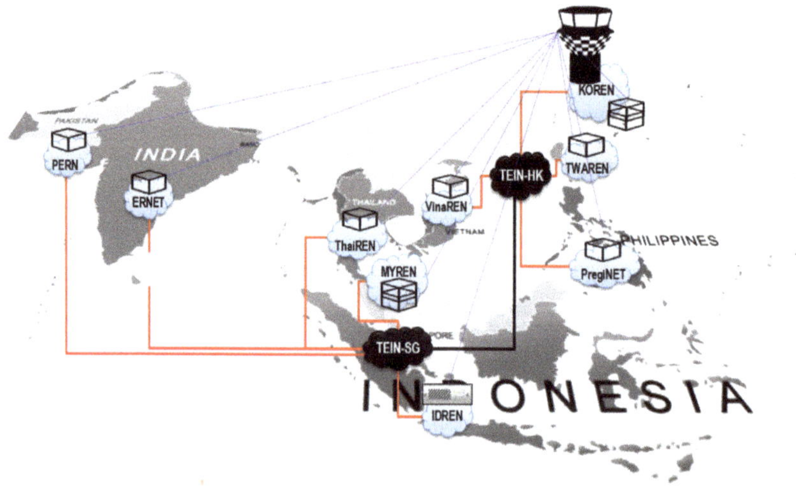

Figure 2. OF@TEIN Playground.

Software-defined Networking (SDN) tools assist developers and operators to prepare the experiment environment in OF@TEIN, by enabling networking resources (e.g., switches and Flowspaces [8]) preparation. Similarly, cloud management software can cover computing resources (e.g., VMs) preparation. Thus, the combination of distributed SDN and cloud testbeds ready to provide scalable and flexible computing resources with enhanced networking capability. However, the seamless integration of SDN-enabled and cloud-leveraged infrastructure is a very challenging task, due to open and conflicting options in configuring and customizing resource pools together. Therefore, we should carefully provision all resource configuration aspects such as multi-site distribution, virtualized resource slicing (i.e., isolation), and multi-tenancy support while considering hardware deployment for SDN and cloud integrated testbed.

As mentioned above, this paper proposes the concept of hyper-converged SmartX Box that can easily accommodate virtualized and programmable resources (i.e., OpenFlow-enabled virtual switches and OpenStack-leveraged cloud VMs) to build the OF@TEIN SDN-enabled multi-site clouds playground (i.e., miniaturized testbed). The collection of OpenFlow-based virtual switches (i.e., OpenvSwitch [9]) is providing SDN capability, which is controlled by both developers and operators SDN controllers. Simultaneously, OpenStack-leveraged VMs (i.e., working as virtual Boxes) are effectively managed by OpenStack cloud management [10]. However, the actual design and implementation of SmartX Box are continuous to evolve for supporting new experiment over OF@TEIN Playground.

1.2. Motivation and Related Work

It is well known that the service-centric networking model to provide higher-level connectivity and policy abstraction is an integral part of cloud-leveraged applications. The emerging SDN paradigm can provide new opportunities to integrate cloud-leveraged services with enhanced networking capability through deeply programmable interfaces and DevOps-style automation. Several SDN-based approaches have been proposed to provide virtualized overlay networking for multi-tenancy cloud infrastructure. For example, Meridian proposed an SDN controller platform to support service-level networking for cloud infrastructure [11]. Similarly, CNG (Cloud Networking Gateway) attempts to address multi-tenancy networking for distributed cloud resources from multiple providers while providing flexibilities in deploying, configuring and instantiating cloud networking services [12]. As a prototype of implementation, the large-scale deployment of GENI Racks over national R&E (research and education) network is also moving towards a programmable, virtualized, and distributed collection of networking/compute/storage resources, a global-scale "deeply programmable cloud". It satisfied research requirements in a wide variety of areas, including cloud-based applications [1]. Another effort from the EU, known as "BonFIRE", is a multi-site testbed that supports testing of cloud-based distributed applications, which offer a unique ease-to-use functionality in terms of configuration, visibility, and control of advanced cloud features for experimentation [13].

Aligned with converged SDI (Software-defined Infrastructure) paradigm [14], the SDN and cloud testbeds should continuously support for the new type of technologies and experiments. To guarantee the usability and continuity of the testbed, the testbed deployment needs to consider the three following aspects. First, the resources should be open without any limitations due to proprietary vendor software or hardware implementation. Second, the resources should be conceptually and reality agile to adopt new technologies or match experiment requirements. Third, the cost and number of physical (i.e., hardware) resources to be deployed should be very minimum for each site of the playground. It only requires a minimized budget (i.e., leveraging low-cost commodity hardware). Please note that leveraging the power of open-source software (e.g., KVM hypervisor [15], LXC Linux container [16], Open virtual switch [9], and others) is also important to support distributed resources centralized management.

Since the cloud computing model is massively adopted for computation infrastructure deployment, it should consider an affordability aspect of the cloud deployment model. In 2017, Ta A.D. [17] tried to address the adoption of a cloud computing framework for developing countries by leveraging hyper-converged servers as virtual computing infrastructure. Moreover, it has gained more acceptance since the networking function (i.e., switch) was extended into a virtualization layer, which allows the creation of multiple virtual switches in the Linux-based server [9]. Then, in 2016, a software-defined framework, called *OpenBox*, tries to address NFs (network functions) deployment by decoupling between NFs control-plane and data-plane that is very similar to SDN solutions [18].

1.3. Aim and Contributions of the Paper

We tried to align with the above testbed deployment, but unfortunately, there are some technical and environment gaps with all those testbed efforts. GENI [1] covers extensive testbed features and functionalities for a different type of experiments while considering increasing the performance with powerful hardware. However, they are not focusing on distributed over heterogeneous infrastructure and the cost of deployment due to the good support of infrastructure and funding sources. FIRE [2] is similar to GENI, but with a less extensible feature because they have much more focus on smaller testbed but also with good performance. They are a little bit concerned about the distributed over heterogeneous infrastructure due to different conditions and policies among many Europeans countries. PlanetLab [19] does not consider features and performance due to the simple requirement of virtual resources with basic networking. However, it is a large-scale distributed testbed because it is deployed over a thousand sites with different network infrastructures. PlanetLab does not concern about the deployment cost because it is supported by a large research community in adding new hardware resources as a new testbed site.

In summary, it is not possible to directly apply their deployment approaches into our environment due to several reasons such as limited funding, level of research interest, and network infrastructure support. Therefore, we want to provide extensive experiment features/functions over a reasonable number of distributed sites over heterogeneous infrastructure. However, at the same time, we need to reduce the number of hardware for each experiment by providing hyper-converged box-style resources. In other words, we want to increase the testbed capability while keeping the cost of the deployment as low as possible. Thus, since 2013, OF@TEIN had considered those mentioned aspects by changing the initial rack-style resource deployment from GENI testbed into box-style, hyper-converged, and server-based resources for an affordable playground across developing countries in Asia region [7]. The main contributions of this paper are:

1. Proposing a concept of an affordable playground with a centralized playground tower and multiple centers to manage and control distributed resources spread across heterogeneous infrastructure.
2. Proposing a design of box-style resources (i.e., SmartX Box), an *open*, *agile* and *economic* hyper-converged resources which able to be deployed and verified over underlying heterogeneous infrastructure.
3. Reducing the cost of the deployment and provisioning time of the playground by leveraging low-cost commodity hardware and developing the open-source-based DevOps automated tools to provision hyper-converged box-style resources.

2. An Affordable Playground

2.1. An SDN-Enabled Multi-Site Clouds Playground

A *Playground* is defined as a miniaturized and customizable testbed that is easy to build and operate for various research experiments by a tiny-size DevOps-style team of developers and operators. Mainly, we focus on establishing a multi-site playground infrastructure where playground resources are physically distributed across multiple geographical sites but logically inter-connected with each other to offer a unified shared pool of resources. A tiny-size team of people (i.e., operators) should provision and control the multi-site isolated resources, which is openly accessible by a group of people (i.e., developers) from all involved sites. As depicted in Figure 3, the proposed multi-site playground has several vital entities, which are discussed below:

1. **Playground Sites with Hyper-converged Boxes**. When playground developers want to perform their experiments, they can dynamically acquire dedicated resources from the pool of multi-site resources. For the customizable (i.e., software-defined) playground, the resource infrastructure of the multi-site playground should be composable (e.g., programmable) in terms of computing, storage, and networking types of resources. By leveraging the growing popularity toward

hyper-converged appliances that integrate computing, storage, and networking resources into a single box-style entity, we can enable the multi-site resource pool that ready to be customized and scaled-out without the manual intervention of playground operators. Thus, in our approach, each playground site is equipped with a box-style of hyper-converged resources, denoted as *SmartX Box*, the hyper-converged box-style resource that should be useful in supporting the required composability by comfortably accommodating virtualized and programmable resources. Multiple types of SmartX Box should be designed and deployed for different purposes, such as SD-WAN, SDN-enabled clouds, and access extension support. However, the critical aspects of those boxes design are similar: *open, agile*, and *economic* resources.

Furthermore, the SmartX Box can be divided into three main abstractions, which are *box, function*, and *inter-connect*. The *box* represents all the server-based hardware that runs Linux as a baseline open-source operating system for other software-based functions. The *function* represents virtual functions (e.g., virtual machine, virtual switch, or virtual router), which are implemented by using a set of open-source software. Finally, the *inter-connect* is representing the path/link between functions or boxes which include the tunnel-based overlay networking because playground sites are spread over the heterogeneous underlay network infrastructure.

2. **Playground Tower with Centers**. To satisfy the dynamic requirements of playground developers on diversified functionalities over distributed but miniaturized resource pools, the proposed playground should integrate the emerging technology paradigms such as the SDN, cloud computing, and Internet of Things (IoT). However, the SDN-enabled multi-site clouds combination brings new complexities for the playground operators, since multi-site clouds resources, connected via SDN-based networks, demand various software-based DevOps-automation tools to build, operate, and use automatically. Thus, we propose the concept of *Playground Tower*, which provides a logical space-like abstraction in a centralized location, which leads the operation of the multi-site playground by following the concept of "monitor and control" tower. From the tower, the DevOps team can enjoy a panoramic view of playground resources that are distributed over underlay networks, and quickly manage and use those resources for their experiments.

The playground tower systematically covers various functional requirements of operating a multi-site playground by employing several centers, also depicted in Figure 3. First, Provisioning Center (P-center) is responsible for remote installation and configuration of multi-site playground resources. Visibility Center (V-center) covers playground visibility and provides panoramic visualization support. Orchestration Center (O-center) handles the management level issues with the assistance of controllers (e.g., SDN/cloud controllers). Thus, several software-based tools are used and developed by leveraging open-source software. For example, to provision SmartX Box, P-Center provides an automation framework such as Chef and MaaS. To continuously operate by re-configuring the playground resources, O-Center provides a set of interfaces (e.g., CLI, API, Web UI) to meet the varying requirements of playground developers. Finally, to monitor the playground resources and traffic flows, V-Center provides playground visibility data in an accessible format for visualization and analysis.

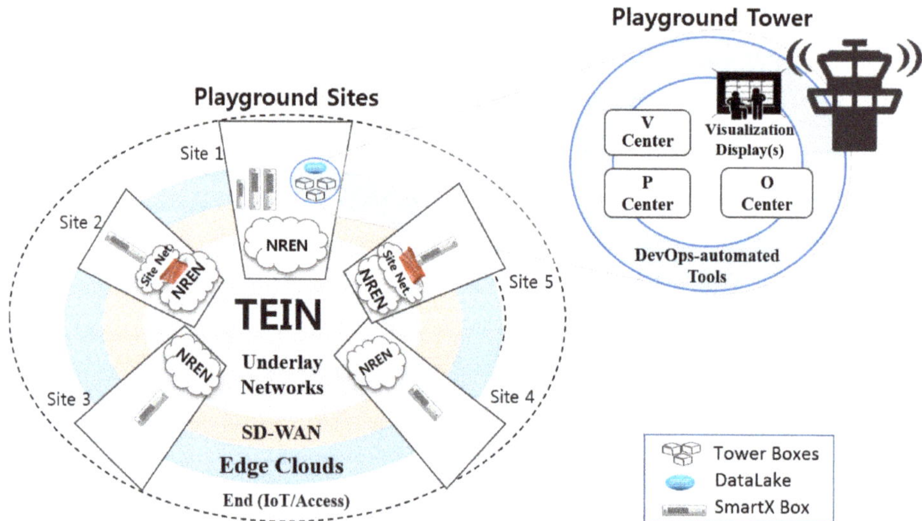

Figure 3. An SDN-Enabled Multi-Site Clouds Playground.

2.2. Affordable Playground with SmartX Box

As mentioned earlier, to specifically address the affordability of multi-site playground, the resources should be open without any configuration/monitoring limitations, agile to adopt new technologies or match experiment requirements, and economics by the deployment of low-cost commodity hardware. Open-source software in converged SDI with SDN, cloud, and Network Function Virtualization (NFV) integration, and the support of open-source hardware are the main drivers of the development of hyper-converged box-style resources.

The concept of box-style hyper-converged resources is shown in Figure 4, which can introduce a unique design of an affordable multi-site playground. The critical design of hyper-converged box-style resource is open, agile, and economic resources. The term box is adopted from a white box (i.e., clear box or open box) that has understandable and controllable subsystems, so it is easy to develop/test software on it without limitation on a vendor-specific feature. The box is open for supporting any server-based hardware because it is fundamentally characterized by software-based components/functions composition and implementation. It is agile to provide a different type of experiment for matching with new upcoming technologies and requirements. It is economical to allow any low-cost commodity server to be used to increase affordability in distributed deployment and operation. However, the actual design of the box is evolved, which include: simplification of the physical hardware specification, components design changes, and modular software implementation.

Unfortunately, provisioning and operating distributed hyper-converged boxes in heterogeneous physical infrastructures quite challenging since it is subjected to different performance parameters (e.g., networks speed and power stability). Also, the independency between multi-domain underlying infrastructure operators (e.g., access and security policies). Consequently, it is hard to maintain the continuous operation of all the boxes. To make those boxes design turnkey simple, we should concentrate on minimizing the requirement and consumed time for provisioning the boxes with these following strategies:

1. *Heterogeneous hardware*: No specific hardware requirements or specific brand/vendor supports, all hardware with acceleration support for virtualization, and multiple network interfaces support for specialized connections can be used.

2. *Remote management*: A distributed deployment and limited access to installation sites are the main reasons to enable the hardware remote management module based on an Intelligent Platform Management Interface (IPMI) [20].
3. *Automated provisioning*: A set of DevOps-based automated provisioning tools is developed to minimize the consumed time for installing and configuring the box with pre-arranged multiple specialized connections.
4. *Centralized configuration template*: The provisioning tool is equipped with a specific template of the configuration file for a specific site of the box that downloadable from a centralized location with a specific component configuration for SDN and cloud.

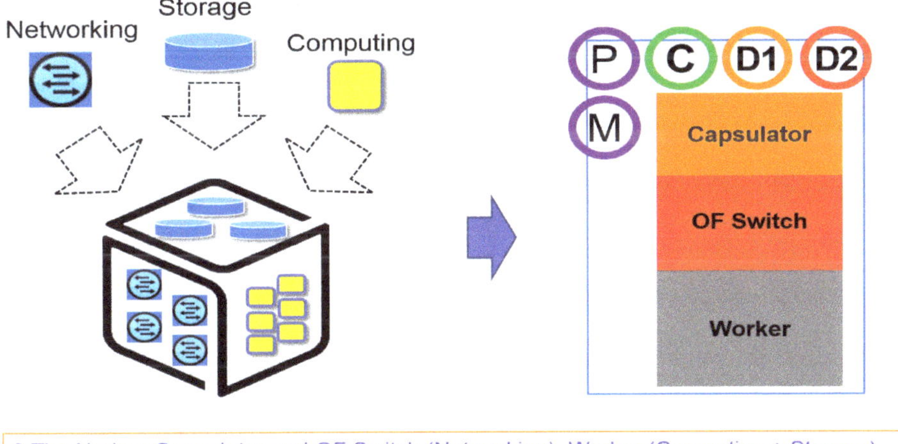

Figure 4. The concept of box-style hyper-converged resources.

2.3. SmartX Box: Design and Implementation

2.3.1. SmartX Box Abstraction

As discussed above, we adopt the hyper-converged box-style resource, called *SmartX Box* [7], as the main building block for our OF@TEIN Playground. The proposed SmartX Box illustrated in Figure 5, as an abstracted format.

The SmartX Box abstraction tries to align the SDN/NFV/cloud integration of SDI with the compute/storage/networking integration of hyper-converged box-style resources. Cloud can provide economics and scalable computing/storage resources without compromising the associated performance, availability, and reliability. SDN provides flexible networking support for highly virtualized computing/storage resources, which is not possible with legacy networking schemes. NFV is assisting SDN by deploying virtualized network functions in the box-style hyper-converged resources of cloud data centers [21]. Virtual monitoring is used to collect box-related data for monitoring and troubleshooting purposes, which one of them is described in our previous work [22]. Moreover, the centralized orchestration for multi-tenant cloud data centers and NFV-assisted SDN infrastructure can provide the simplified orchestration of SDI-ready playground with the slice-based network virtualization support. Thus, by considering the above features, SmartX Box is designed to be ready for supporting a wide range of research experiments. Also, by merging all the required functionalities into the hyper-converged SmartX Box, it is easier to realize the scale-out capability of the playground by simply adding hyper-converged SmartX Boxes to increase the resource capacity of the playground.

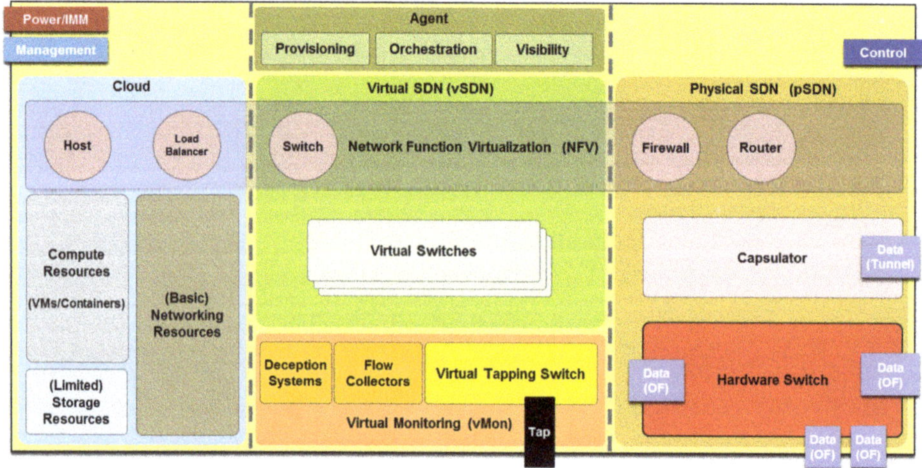

Figure 5. Hyper-converged SmartX Box: Abstraction to match SDN/NFV/Cloud integration.

To support the flexible remote configuration, each hyper-converged SmartX Box requires dedicated and specialized connections for P/M/C/D (power, management, control and data), which are explained in the next section [7]. However, besides those connections, there are no specific hardware requirements for hyper-converged SmartX Boxes. Therefore, any commodity hardware with reasonable computing, storage, and networking resources can be used. The total amount of hardware resources only affects the capacity (e.g., the total number of VM instances per flavor types) in specific boxes, sites, and regions. However, it is essential to consider the hardware acceleration support for virtualization and networking.

2.3.2. Virtualized SDN-Enabled Switches and Cloud-Leveraged VMs

As described above, the conversion from SmartX Racks to SmartX Boxes are completed to manage better the distributed multi-site cloud-based services on the top of SDN-enabled inter-connect capabilities [7]. Thus, the actual design of SmartX Box needs to consider and balance both cloud and SDN aspects carefully. The OpenStack [10] Cloud can provide VM instances and basic networking options for diverse tenants. For SDN, several instances of virtual switches (based on Open vSwitch [9]) are provisioned while allowing users/developers to share them simultaneously. We arrange the SDN and cloud relevant functions inside a single hyper-converged SmartX Box, as shown in Figure 6. SDN-related virtual functions consist of several virtual switches with different roles, e.g., creating developers networking topology, inter-connecting OpenFlow-based overlay networking, and tapping flows for troubleshooting. Also, cloud-related functions are placed to include VM instances for cloud-based applications and to support external connections to VMs.

The inside view of SDN-/cloud-related functions is depicted in Figure 7. First, several SDN-enabled virtual switches are placed and matched with its functionalities: *brcap* for capsulator (encapsulate OpenFlow packets through an overlay tunnel), *br1* and *br2* for users/developers switches, and *brtap* for tapping purpose (capturing packets for troubleshooting as described in [11]). Cloud-related VM instances (a.k.a., virtual Boxes: vBoxes) are managed by KVM hypervisors, which is controlled by OpenStack Nova with specific flavors and images. Additionally, virtual switches (i.e., *br-int*, *br-ex*, and *br-vlan*) and user-space virtual router are configured by OpenStack Neutron to provide required connectivity to cloud VM instances.

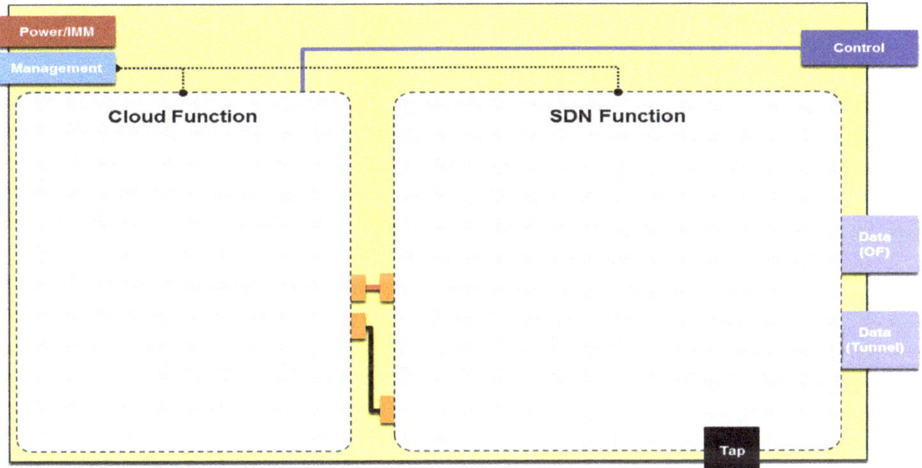

Figure 6. Hyper-converged SmartX Box Design for SDN and Cloud Integration.

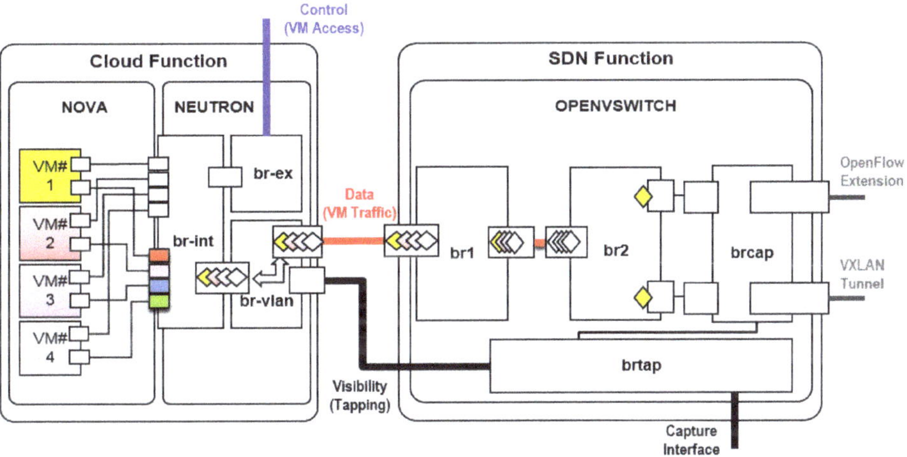

Figure 7. Hyper-converged SmartX Box Virtualized Cloud and SDN Components.

2.4. Semi-automated Resource Provisioning

Deploying hyper-converged SmartX Boxes in heterogeneous physical (i.e., network topology) infrastructures is very troublesome since it is subject to different performance parameters and independent network administrative domains. As mentioned in the previous section, it is tough to keep the sustainable operation of all the boxes. Thus, a set of automated provisioning tools is developed to minimize the consumed time for provisioning (i.e., installing and configuring) all the SmartX Boxes with pre-arranged P/M/C/D connections. The P (Power) connection is used for power up/down SmartX Box. The M (Management) connection is mainly used for managing SmartX Box by the operator. Also, the C (Control) connection is used to access and control the SDN-/Cloud-related functions (i.e., virtual switches and VMs) by the developer. Finally, the D (Data) connection is used for any data-plane traffic that includes inter-connection traffic among multiple SmartX Boxes. Also, the automated provisioning tools are controlled by a centralized P-Center inside the Playground Tower, which has full access to all distributed hyper-converged SmartX Boxes.

First, to automate the provisioning of the SDN-enabled virtual switches, *ovs-vsctl* high-level management interface for OpenvSwitch is used. Please note that *ovsdb* (OpenvSwitch database) protocol is also used for the centralized configuration of the OpenvSwitch database inside each SmartX Box. The provisioning task includes the creation of virtual switches, the configuration of virtual ports/links and overlay tunnel inter-connections, and the control connection of virtual switches and SDN controllers. Next, open-source OpenStack cloud software has special installation and configuration tools, called *DevStack*, which can support several modes of OpenStack configurations with selected operating systems (e.g., Ubuntu, Redhat Enterprise Linux, and CentOS) [23]. For OF@TEIN Playground, we customize DevStack provisioning template to facilitate multi-regional OpenStack cloud deployment with centralized management and authentication.

The overall implementation of semi-automated provisioning for SDN/cloud-enabled SmartX Box is depicted in Figure 8. It is started with a clean-up of previous software installation and checking/upgrading the operating system. Then it is followed with box installation to install/configure OpenStack cloud and OVS components, and finally API-based tools to verify function installation/configuration.

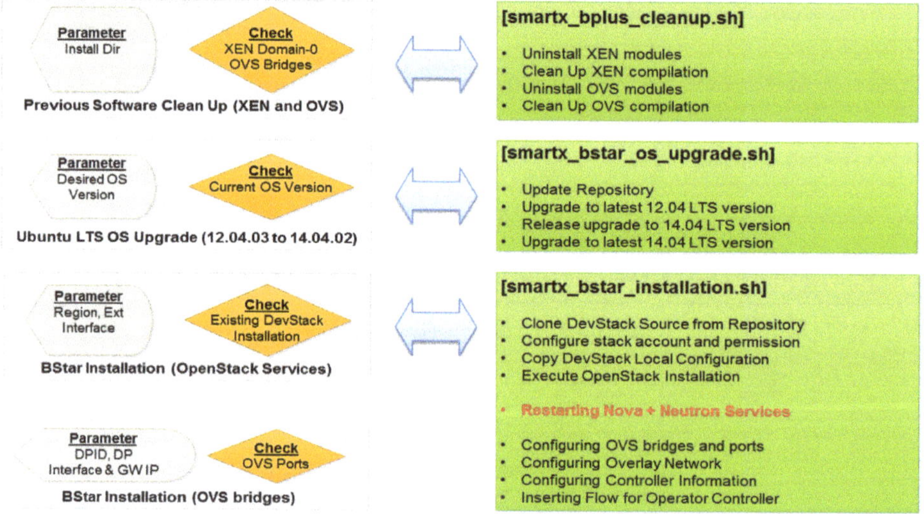

Figure 8. Implementation of Semi-automated Provisioning for SDN/cloud-enabled Hyper-converged SmartX Box.

2.5. SDN and Cloud Centralized Control

Cloud-related virtual functions are inter-connected through SDN-related virtual functions to provide end-to-end communication for multi-site cloud-based applications. The cloud-related virtual functions are controlled centrally by open-source *OpenStack* Cloud management and orchestration software [10]. The SDN-related virtual functions are also centrally controlled by open-source SDN controller such as ODL (Open Daylight) [24] and ONOS (Open Network Operating System) SDN Controller [25]. OpenStack *Keystone* provides centralized user authentication and authorization. OpenStack *Nova* and OpenStack *Neutron* can create VM instances and to provide enhanced connectivity, respectively. The SDN controller (i.e., ODL) manipulates the flow table entries of SDN-enabled virtual switches to enable the flexible steering of inter-connection flows among various functions located in different cloud sites. Both cloud and SDN control software are required to mix and match the configurations so that we can ensure the consistent connections between cloud VM instances. Remember that the main challenge is how to accommodate cloud-based multi-tenancy virtual networks (e.g., flat, VLAN, or tunneled network) for OpenFlow-based network slicing (e.g., IP subnets, VLAN IDs, and TCP/UDP ports). Eventually, VLAN-based multi-tenancy traffic control (e.g., tagging, steering, and mapping) is chosen to integrate tenant-based and sliced-based networking in SDN-enabled multi-site clouds playground.

By manipulating both OpenStack and ODL SDN controller, a VLAN-based multi-tenancy traffic control is implemented as follows. First, we place VMs in two cloud regions and prepare the connectivity for these VMs. These VMs are tagged by OpenStack Nova with a specific tag ID. Second, OpenStack Neutron automatically maps the tag into VLAN ID that matched with SDN-based slice parameters. This matching allows inter-connection flows for VMs to be steered by the developer's SDN controller, supervised by *FlowVisor* [8]. The SDN-based flow steering inserts flow table entries according to the particular incoming and outgoing ports in the developer's virtual switches, where several ports are mapped to other cloud regions/sites. Finally, based on the destination site, it maps to a specific tunnel interface that is pre-configured by the SDN controller of operators. The example control of SDN and cloud components inside the SmartX Box for connecting several VMs from different tenants across multiple sites, as depicted in Figure 9.

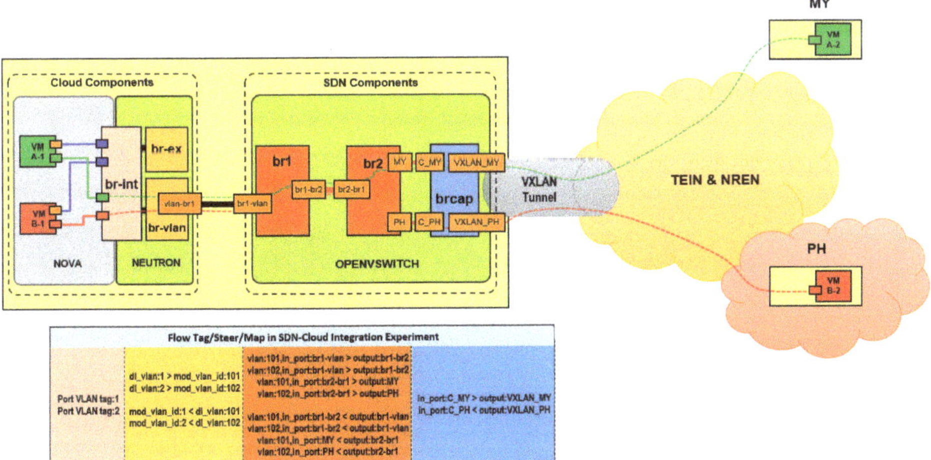

Figure 9. SDN and Cloud Control (Tag, Steer and Map) inside the SmartX Box.

3. Cost and Efficiency Analysis

3.1. TCO Analysis of Conversion from SmartX Rack into SmartX Box

To estimate the reduction of TCO (Total Cost of Ownership) from SDI based on the hyper-converged resources, we can observe these two reports. First, typical three-year server TCO from IDC (International Data Corporation) [26] in 2017, which defined the composition of several costs such as hardware, software, staff training, outsourced cost, user productivity, and staffing (manpower), as depicted as Figure 10a. Second, the value of SDI to reduce the total TCO of 10,000 OS instances which is released by Intel [27], as depicted in Figure 10b. Where manpower efficiency improves up to 60%, software savings up to 70%, hardware reduction up to 20%, and other reduction (infrastructure and energy) decreases up to 20%.

Based on both of the analysis reports, we can estimate the total saving of our conversion effort from SmartX Rack into SmartX Box hyper-converged resources. In summary, as depicted in Figure 11a, the TCO for SDI-ready box-style hyper-converged resources is described as follows. The hardware-related project cost reduced to 5.6% due to single box type of deployment, the software cost decrease to 2.1% due to open-source software adoption, and the most important is the cost for playground operators go down till 36%. The overall TCO saving of SmartX Rack to SmartX Box conversion is around 30.3%. If the graph is normalized into 100% of the pie chart, so we can produce the chart as depicted in Figure 11b. In conclusion, with the same amount of project budget, more cost can be allocated for staff training (e.g., more effort for the research) to increase the quality of operators/developers and also consider more compensation on the experiment downtime or the developers/researchers productivity.

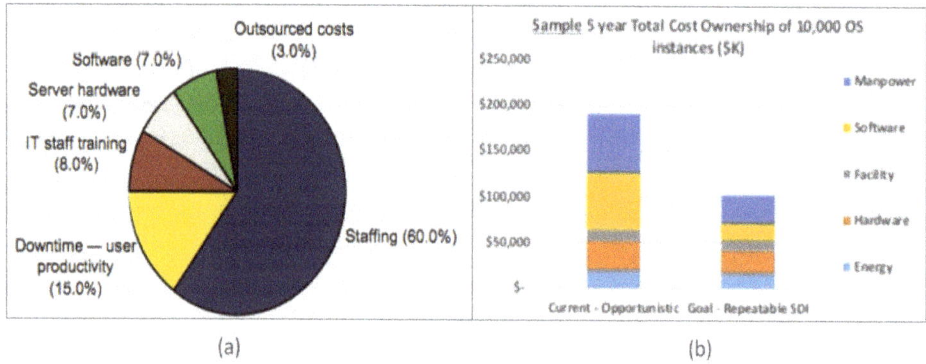

(a) (b)

Figure 10. (**a**) Server-based 3-years TCO Composition [26] and (**b**) SDI Value to reduce 5 years TCO [27].

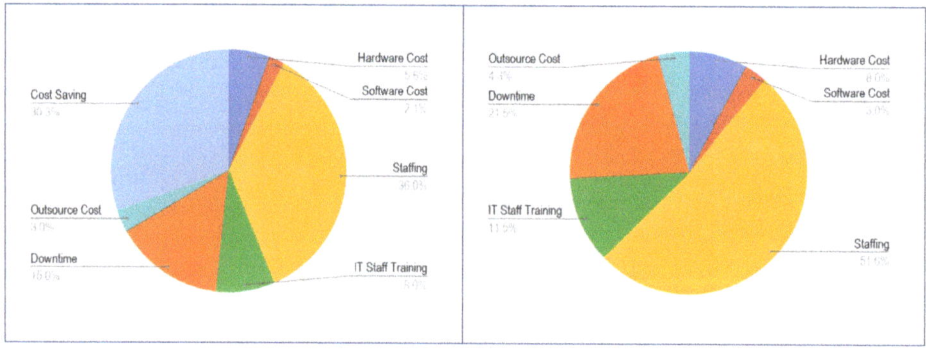

Figure 11. (**a**) TCO saving with hyper-converged infrastructure and (**b**) Normalized TCO with hyper-converged infrastructure.

3.2. The Efficiency of Semi-Automated Provisioning

To facilitate the agile deployment of OF@TEIN Playground, both SDN-/cloud-related tools are used for automated provisioning of hyper-converged SmartX Boxes. It is aligned with the recent employment of DevOps automation since the OF@TEIN Playground is operated by a limited number of operators and becomes easily uncontrollable as it spans across multi-domain inter-connected networks beyond the privileges of playground operators. Thus, by using DevStack-based OpenStack deployment and ovs-vsctl or ovsdb protocol for virtual switch provisioning, we can simplify the semi-automated provisioning of hyper-converged SmartX Boxes. Also, REST APIs of the ODL SDN controller is used for automated flow insertion, flow modification, and flow deletion. In summary, most of the provisioning steps are automated, except for manual handling of critical tasks such as DevStack-based OpenStack service restart and VXLAN tunnel checking/recovery.

The duration of the whole process of semi-automated provisioning depends on the Internet connection speed due to the online OS upgrade and online OpenStack software installation. However, it takes less time for new box clean installation (including box restart), because previous software clean-up and OS upgrade is not required. Moreover, re-configuring pre-installed SmartX Box is much faster with "offline mode" enabled because online software/package copy from Ubuntu and OpenStack repositories are not required. Figure 12a shows the semi-automated provisioning results, which take approximately 50 min for fully upgrading a SmartX Box with network connection up to 300 Mbps. Thus, it takes around 6 h for the slowest network connection which less than 10 Mbps. However, it takes only around 20 min (including box restart) for provisioning without cleaning up the previous installation and upgrading the operating system. While Figure 12b justifies the longest completion time is for upgrading the operating systems and restarting the SmartX Box. The clean-up task is negligible because it is less than one minute, and then the installation task is reasonable for such a customized configuration. Moreover, it takes less than 10 min to recover or re-configure SmartX Box with "offline mode" [7].

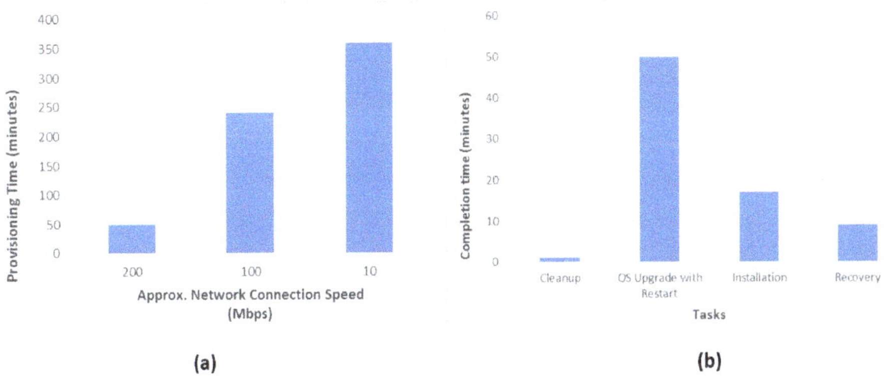

Figure 12. Semi-automated Remote Provisioning Result for SmartX Box: (**a**) Provisioning time for the different network connection speed, and (**b**) Completion time for different provisioning task.

4. Playground Deployment Verification

4.1. Distributed Deployment of SmartX Boxes for Building OF@TEIN Multi-Site Playground

Multiple hyper-converged SmartX Boxes are deployed on existing hardware of OF@TEIN Playground with a special focus on adding the open-source OpenStack cloud-management software. The OF@TEIN Playground relies on the heterogeneous physical underlay infrastructure across multiple administrative domains. Thus, the multi-regional OpenStack cloud deployment is currently investigated as the deployment option because it gives simple and common configuration for all regions (i.e., SmartX Box sites). It also supports an independent IP addressing scheme and has less dependency on the overlay networking among regions. Despite multi-regional independent cloud deployment, the OF@TEIN Playground supports an integrated cloud management interface by deploying web-based OpenStack Horizon UI and the centralized account/token authentication from OpenStack Keystone. The resulting OpenStack multi-regional cloud deployment is illustrated in Figure 13.

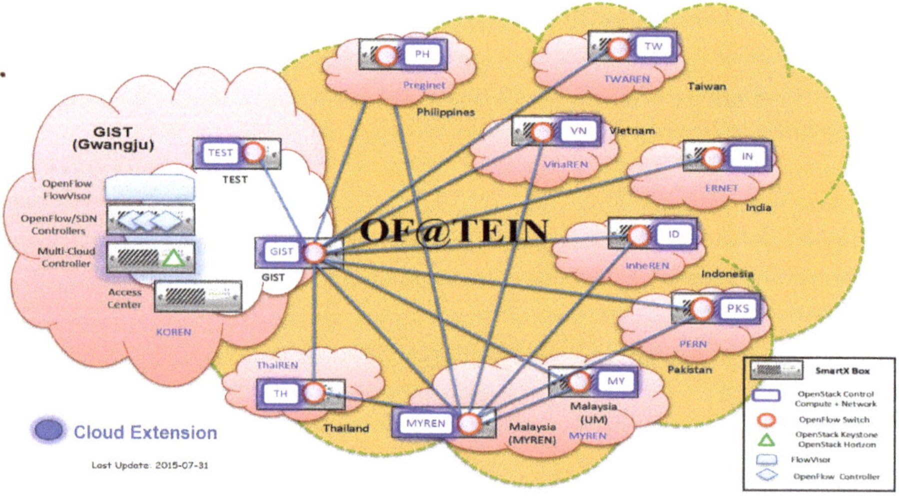

Figure 13. OpenStack Multi-Regional Configuration in OF@TEIN Playground.

Next, the OF@TEIN Playground is enhanced with multiple mesh-style inter-connections of NVGRE/VXLAN overlay tunnels, along with a unique flow-tapping virtual switch [22]. The OpenStack multi-region deployment is modified to build an SDN-enabled multi-site playground where inter-VM connectivities between cloud VMs are used by leveraging OpenFlow-enabled data planes. The data planes are programmed and controlled by the centralized SDN controller, co-located with centralized cloud management.

4.2. Example of Experiment with SDN-Enabled Multi-Site Cloud Playground

This example shows both aspects of experiment preparation in SDN-enabled multi-site clouds playground to provision resources in both OpenStack and ODL SDN controller. First, a VM in one of the cloud regions (i.e., playground sites) is prepared, including the basic connectivity for this VM using the first VM virtual NIC (Network Interface Card). It is connected to a control network, called "private", for providing VM remote access from an external network. Then, second VM virtual NIC is connected to a data network, called "datapath01", which is automatically mapped into pre-configured VLAN ID that matched with SDN-based slice parameters. This mapping allows the flow from this VM to be steered by the developer's SDN controller, supervised by FlowVisor [8]. Another VM in the second region also is prepared with the same steps as depicted in Figure 14. Next, the SDN-based flow steering that is leveraging the ODL SDN controller inserts flow table entries in developers' virtual switches from/to those VMs to/from pre-defined ports that are already mapped to other cloud regions/sites. Those pre-defined ports are mapped to a tunnel interface from the originating site into the designating site that is controlled by the operators' SDN controller of operators. The example of the steps is depicted in Figure 15.

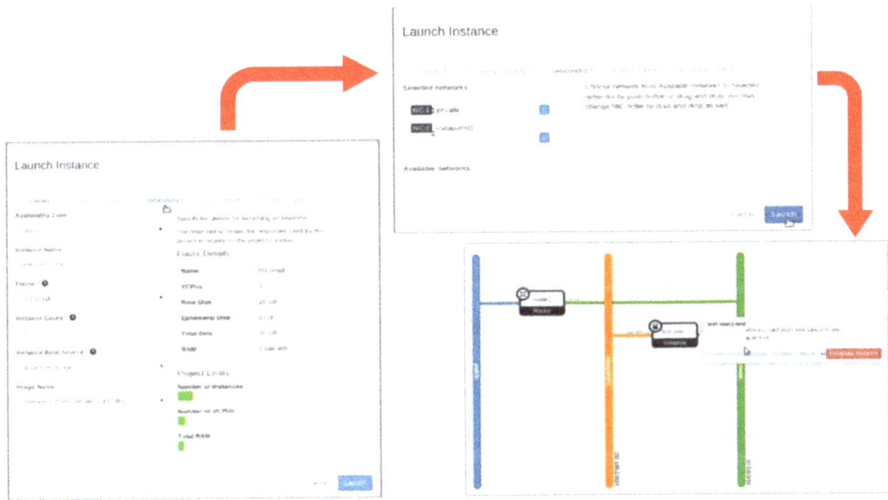

Figure 14. OpenStack VMs Preparation and Configuration.

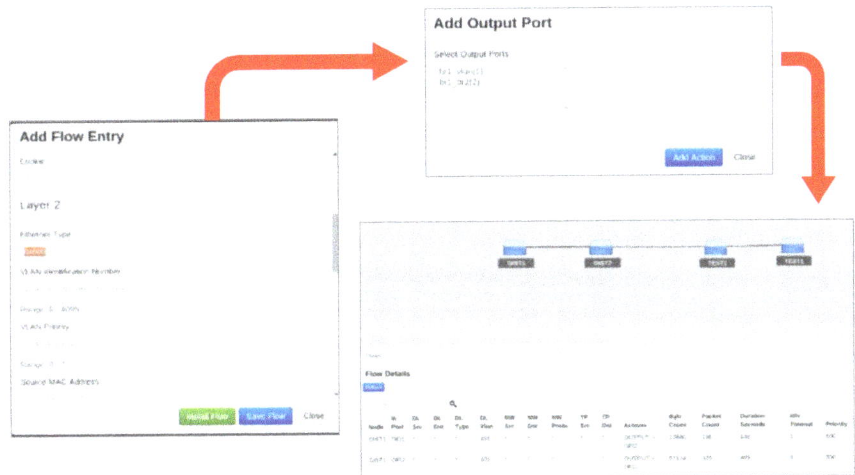

Figure 15. Flow Configuration in OpenDaylight Controller.

4.3. Multi-site Playground Visibility and Visualization

Resource-level monitoring and visualization are important operation activities for OF@TEIN playground. The SDN and cloud components implementation for hyper-converged SmartX Boxes provides diverse physical and virtualized resource combinations while at the same time brings new complexities for monitoring and visualization. The SDN-cloud-enabled playground demands considerably different resource-level visibility solutions from traditional networking testbed. A distinctive, component-based, data-oriented approach is required for resource-level visibility of distributed OF@TEIN physical and virtual resources. By integrating open-source software/tools, we set up a unique resource-level visibility solution, which is focused on operation data collection from multiple sources and interactive large-scale visualization [28]. Resource-level visibility data is collected in nearly real-time to help the operators for monitoring the status of the resources by using a single and unified visibility user interface. Large-scale visualization (i.e., network tiled display leverages SAGE Framework [29]) enables the simultaneous visualization of the multiple and different types of visualization sources (e.g., web-based UI, remote desktop, secure shell), as depicted in Figure 16. It allows OF@TEIN operators to manage the resources and developers to execute the experiment while keeping an eye on the playground resources.

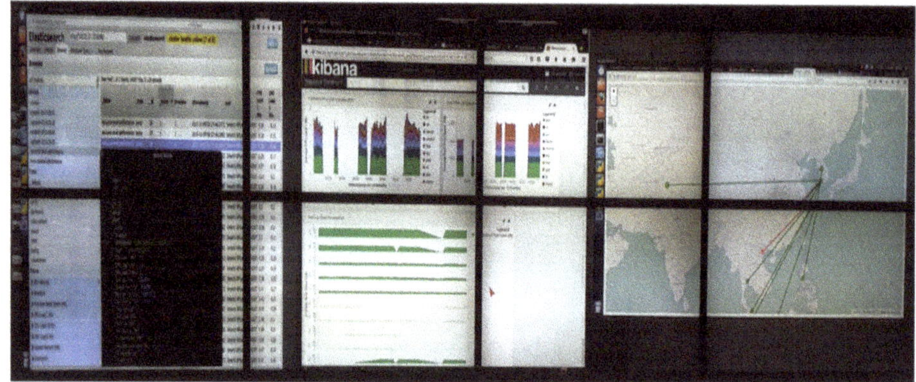

Figure 16. Multi-site playground resources visibility over network tiled display.

5. Conclusions

This paper gives comprehensive discussions of the unique concept and design of SDN-enabled multi-site clouds playground with hyper-converged box-style resources for an innovative and diverse research experiment. OF@TEIN Playground is successfully provisioned as an affordable SDN-enabled multi-site clouds playground with distributed SmartX Boxes deployment for integrated SDN and cloud experiments. We believe that an open, agile, and economics box-style hyper-converged resources can provide a larger scale of an affordable and sustainable playground for diverse experiments with a wide variety of application areas.

Author Contributions: Conceptualization—J.K. and A.C.R., Supervision—J.K., Investigation—A.C.R., Software and Validation—A.C.R., and M.U., Visualization—M.U., Writing—A.C.R., and Reviewing and Editing—J.K. and M.U.

Funding: This work was supported by Institute of Information & Communications Technology Planning & Evaluation (IITP) grant of the Korea Government (MSIT) (No. 2015-0-00575, Global SDN/NFV Open-Source Software Core Module/Function Development, and No. 2017-0-00421, Cyber Security Defense Cycle Mechanism for New Security Threats). This research is also partially supported by Asi@Connect grant of the Asi@Connect-17-094 (No. IF050-2017), OF@TEIN+: Open/Federated Playground for Future Networks.

Acknowledgments: The authors would like to acknowledge gratefully to OF@TEIN Community who has been supported the deployment and operation of OF@TEIN playground for the past few years. Especially for playground operators and NREN administrators who help us to maintain the stability and connectivity of the resources to be able continuously used by playground developers/users. Hopefully, the collaboration amongst all the involved parties can be continued to enhance and extend our proposed concept and effort in the next few years.

Conflicts of Interest: The authors declare no conflict of interest.

Abbreviations

The following abbreviations are used in this manuscript:

SDN	Software-defined Networking
GENI	Global Environment for Network Innovations
FIRE	Future Internet Research and Experimentation
OF@TEIN	OpenFlow at Trans-Eurasia Information Network
DevOps	Development and Operations
LAN	Local Area Network
VM	Virtual Machine

CNG   Cloud Networking Gateway
SDI   Software-defined Infrastructure
KVM   Kernel Virtual Machine
LXC   Linux Container
OVS   OpenvSwitch
NF   Network Function
SD-WAN   Software-defined Wide-area Network
IoT   Internet of Things
MaaS   Machine as a Service
API   Application Programming Interface
CLI   Command Line Interface
NFV   Network Function Virtualization
IPMI   Intelligent Platform Management Systems
ODL   Open Daylight
ONOS   Open Networking Operating System
VLAN   Virtual LAN
TCP   Transport Control Protocol
UDP   User Datagram Protocol
TCO   Total Cost Ownership
IDC   International Data Corporation
NVGRE   Network Virtualization using Generic Routing Encapsulation
VXLAN   Virtual Extensible LAN
REST   Representational State Transfer
SAGE   Scalable Adaptive Graphics Environment

References

1. Berman, M.; Chase, J.S.; Landweber, L.; Nakao, A.; Ott, M.; Raychaudhuri, D.; Ricci, R.; Seskar, I. GENI: A federated testbed for innovative network experiments. *Comput. Netw.* **2014**, *61*, 5–23. [CrossRef]
2. Gavras, A.; Karila, A.; Fdida, S.; May, M.; Potts, M. Future Internet Research and Experimentation: The FIRE Initiative. *SIGCOMM Comput. Commun. Rev.* **2007**, *37*, 89–92. [CrossRef]
3. Kim, J.; Cha, B.; Kim, J.; Kim, N.L.; Noh, G.; Jang, Y.; An, H.G.; Park, H.; Hong, J.; Jang, D.; et al. OF@ TEIN: An OpenFlow-enabled SDN testbed over international SmartX Rack sites. *Proc.-Asia-Pac. Adv. Netw.* **2013**, *36*, 17–22. [CrossRef]
4. Loukides, M. *What is DevOps*? O'Reilly Media, Inc.: Newton, MA, USA, 2012.
5. Marschall, M. *Chef Infrastructure Automation Cookbook*; Packt Publishing Ltd.: Birmingham, UK, 2015.
6. Na, T.; Kim, J. Inter-connection automation for OF@ TEIN multi-point international OpenFlow islands. In Proceedings of the Ninth International Conference on Future Internet Technologies, Nice, France, 24–28 July 2016; p. 12.
7. Risdianto, A.C.; Shin, J.; Kim, J. Building and Operating Distributed SDN-Cloud Testbed with Hyper-convergent SmartX Boxes. In Proceedings of the International Conference on Cloud Computing, Nice, France, 22–27 March 2015; pp. 224–233.
8. Sherwood, R.; Gibb, G.; Yap, K.K.; Appenzeller, G.; Casado, M.; McKeown, N.; Parulkar, G. *Flowvisor: A Network Virtualization Layer*; OpenFlow Switch Consortium, Tech. Rep.; 2009; Volume 1, p. 132. Available online: https://pdfs.semanticscholar.org/64f3/a81fff495ac336dccdd63136d451852eb1c9.pdf (accessed on 30 October 2019).
9. Pfaff, B.; Davie, B. *The Open vSwitch Database Management Protocol*; Technical Report, RFC 7047; IETF: Fremont, CA, USA, 2013.
10. Sefraoui, O.; Aissaoui, M.; Eleuldj, M. OpenStack: Toward an open-source solution for cloud computing. *Int. J. Comput. Appl.* **2012**, *55*, 38–42. [CrossRef]
11. Banikazemi, M.; Olshefski, D.; Shaikh, A.; Tracey, J.; Wang, G. Meridian: An SDN platform for cloud network services. *IEEE Commun. Mag.* **2013**, *51*, 120–127. [CrossRef]
12. Mechtri, M.; Houidi, I.; Louati, W.; Zeghlache, D. SDN for inter cloud networking. In Proceedings of the 2013 IEEE SDN for Future Networks and Services (SDN4FNS), Trento, Italy, 11–13 November 2013; pp. 1–7.

13. Hume, A.C.; Al-Hazmi, Y.; Belter, B.; Campowsky, K.; Carril, L.M.; Carrozzo, G.; Engen, V.; García-Pérez, D.; Ponsatí, J.J.; Kűbert, R.; et al. Bonfire: A multi-cloud test facility for internet of services experimentation. In Proceedings of the International Conference on Testbeds and Research Infrastructures, Thessanoliki, Greece, 11–13 June 2012; pp. 81–96.

14. Kang, J.M.; Lin, T.; Bannazadeh, H.; Leon-Garcia, A. Software-Defined Infrastructure and the SAVI Testbed. In *Testbeds and Research Infrastructure: Development of Networks and Communities*; Leung, V.C., Chen, M., Wan, J., Zhang, Y., Eds.; Springer International Publishing: Cham, Switzerland, 2014; pp. 3–13.

15. Kivity, A.; Kamay, Y.; Laor, D.; Lublin, U.; Liguori, A. kvm: The Linux virtual machine monitor. *Linux Symp.* **2007**, *1*, 225–230.

16. Helsley, M. LXC: Linux Container Tools. 2009; p. 11. Available online: https://developer.ibm.com/tutorials/l-lxc-containers/ (accessed on 30 October 2019)

17. Ta, A.D. A Cloud Computing Adoption Framework for Developing Countries. In *Sustainable ICT Adoption and Integration for Socio-Economic Development*; IGI Global: Hershey, PA, USA, 2017; pp. 175–199.

18. Bremler-Barr, A.; Harchol, Y.; Hay, D. OpenBox: A software-defined framework for developing, deploying, and managing network functions. In Proceedings of the 2016 ACM SIGCOMM Conference, Florianopolis, Brazil, 22–26 August 2016; pp. 511–524.

19. Chun, B.; Culler, D.; Roscoe, T.; Bavier, A.; Peterson, L.; Wawrzoniak, M.; Bowman, M. Planetlab: An overlay testbed for broad-coverage services. *ACM SIGCOMM Comput. Commun. Rev.* **2003**, *33*, 3–12. [CrossRef]

20. Intelligent Platform Management Interface. Available online: https://www.intel.com/content/www/us/en/servers/ipmi/ipmi-home.html (accessed on 5 October 2019).

21. ETSI GS NFV *001: "Network Functions Virtualisation (NFV)"*; Architectural Framework; 2013. Available online: https://www.etsi.org/deliver/etsi_gs/NFV/001_099/002/01.01.01_60/gs_NFV002v010101p.pdf (accessed on 30 October 2019)

22. Risdianto, A.C.; Kim, J. Flow-centric visibility tools for OF@ TEIN OpenFlow-based SDN testbed. In Proceedings of the 10th International Conference on Future Internet, Seoul, Korea, 8–10 June 2015; pp. 46–50.

23. DevStack. Available online: https://docs.openstack.org/devstack/latest/ (accessed on 19 September 2019).

24. Medved, J.; Varga, R.; Tkacik, A.; Gray, K. Opendaylight: Towards a model-driven sdn controller architecture. In Proceedings of the 2014 IEEE 15th International Symposium on World of Wireless, Mobile and Multimedia Networks (WoWMoM), Coimbra, Portugal, 21–24 June 2016; pp. 1–6.

25. Berde, P.; Gerola, M.; Hart, J.; Higuchi, Y.; Kobayashi, M.; Koide, T.; Lantz, B.; O'Connor, B.; Radoslavov, P.; Snow, W.; et al. ONOS: Towards an open, distributed SDN OS. In Proceedings of the Third Workshop on Hot Topics in Software Defined Networking, Chicago, IL, USA, 17–22 August 2014; pp. 1–6.

26. Perry, R.; Gillen, A. *Demonstrating Business Value: Selling to Your C-Level Executives*; Technical Report; International Data Corporation (IDC): Framingham, MA, USA, 2007.

27. *Value of SDI*; Intel IT TCO Tools; Intel Corporation. Available online: https://www.slideshare.net/UOLDIVEO1/intel-05-razes-para-adotar-openstack-na-sua-estratgia-de-cloud (accessed on 30 October 2019).

28. Usman, M.; Risdianto, A.C.; Kim, J. Resource monitoring and visualization for OF@ TEIN SDN-enabled multi-site cloud. In Proceedings of the IEEE 2016 International Conference on Information Networking (ICOIN), Jeju Island, Korea, 31 January–2 February 2005; pp. 427–429.

29. Renambot, L.; Rao, A.; Singh, R.; Jeong, B.; Krishnaprasad, N.; Vishwanath, V.; Chandrasekhar, V.; Schwarz, N.; Spale, A.; Zhang, C.; et al. Sage: The scalable adaptive graphics environment. In Proceedings of the WACE, Nice, France, 23–24 September 2004; Volume 9, pp. 2004–2009. Available online: https://www.evl.uic.edu/documents/sage_wace2004.pdf (accessed on 30 October 2019).

Article

Phantom: Towards Vendor-Agnostic Resource Consolidation in Cloud Environments [†]

Aaqif Afzaal Abbasi [1], Mohammed A. A. Al-qaness [2], Mohamed Abd Elaziz [3], Ammar Hawbani [4], Ahmed A. Ewees [5], Sameen Javed [6] and Sunghwan Kim [7,*]

[1] Department of Software Engineering, Foundation University, Islamabad 44000, Pakistan; aaqif.afzaal@fui.edu.pk

[2] School of Computer Science, Wuhan University, Wuhan 430072, China; alqaness@whu.edu.cn

[3] Department of Mathematics, Faculty of Science, Zagazig University, Zagazig 44519, Egypt; abd_el_aziz_m@yahoo.com

[4] School of Computer Science and Technology, University of Science and Technology of China, Hefei 230072, China; anmande@ustc.edu.cn

[5] Department of Computer, Damietta University, Damietta 34511, Egypt; ewees@du.edu.eg

[6] Quality Assurance Department, Keeptruckin Inc., Islamabad 44000, Pakistan; sameen.javed@keeptruckin.com

[7] School of Electrical Engineering, University of Ulsan, Ulsan 44610, Korea

[*] Correspondence: sungkim@ulsan.ac.kr; Tel.: +82-52-259-1401

[†] This research is an extended version of a paper published in Proceedings of the 3rd IEEE International Conference on Data Science and Computational Intelligence (DSCI'19), Shenyang, China, 21–23 October 2019.

Received: 24 September 2019; Accepted: 15 October 2019; Published: 18 October 2019

Abstract: Mobile-oriented internet technologies such as mobile cloud computing are gaining wider popularity in the IT industry. These technologies are aimed at improving the user internet usage experience by employing state-of-the-art technologies or their combination. One of the most important parts of modern mobile-oriented future internet is cloud computing. Modern mobile devices use cloud computing technology to host, share and store data on the network. This helps mobile users to avail different internet services in a simple, cost-effective and easy way. In this paper, we shall discuss the issues in mobile cloud resource management followed by a vendor-agnostic resource consolidation approach named Phantom, to improve the resource allocation challenges in mobile cloud environments. The proposed scheme exploits software-defined networks (SDNs) to introduce vendor-agnostic concept and utilizes a graph-theoretic approach to achieve its objectives. Simulation results demonstrate the efficiency of our proposed approach in improving application service response time.

Keywords: cloud computing; management; middle box; placement; resource; SDN; vendor-agnostic; virtual machine; VM

---

1. Introduction

Mobile-oriented future networks [1–3] are gaining tremendous importance in the field of computing and networking industry. With the advent of wireless networking technologies, the wide-scale use of smartphone devices and the World Wide Web is being shifted rapidly from static to mobility-based solutions. For example, mobile service users will exceed two billion users [4,5]. Such drastic changes are influencing the way IT concepts used to act and behave in the past.

However, the original idea of the internet was not based on mobility-based services. In other words, it can be said that the original idea of the internet was meant for fixed hosts instead of mobile hosts. So with the emergence of mobile technology, various patch-on protocols were introduced to support mobile environments e.g., Mobile IP [6,7] and its variants. However, patch-on technology based solutions also have their limitations.

In an environment based on mobile computing, support for mobility is a vital requirement rather as an add-on feature. The legacy network protocolsmainly focus on fixed hosts. In terms of usability, legacy protocols often describe mobility as an additional functionality of a device. This behavior leads to the creation of protocols based on mobility and is related to the modified versions of TCP/IP protocols suite [8–10]. With these trends of mobility-awareness in the protocols, unexpected degradation of performance, such as overuse of proxy, triangle routing, etc., is induced.

Mobile devices, including tablet PC or smartphones, are increasingly becoming an important part of our lives as a vital and easy sources of communication tools that are not bounded by the elements of time and space [11]. Mobile users utilize multiple mobile-based services by using different kinds of mobile apps. These apps are hosted on remote servers through wireless networks. The fast growth witnessed in mobile computing is a very prominent factor in the IT industry. It also influenced the commerce industry. However, with this fast growth of mobile devices, we are also facing numerous challenges such as computing resources car city, bandwidth allocation, storage and retrieval challenges and battery life time. Therefore, it can be safely said that the limitations of computing resources greatly hinder the betterment of computing services quality.

Cloud computing has been accepted as the infrastructure of next-generation networks [12]. Cloud users can benefit through cloud infrastructure by using various services (such as storage and services hosting), platforms (operating systems, middleware and related services) and software (applications) supported by cloud-enabled services like Amazon, Salesforce or Google at low prices. Furthermore, cloud computing enables its users to broadly utilize the resources on a pay-per-use policy [13]. By using such mobile applications, users can benefit from various cloud computing functions. With the rapid growth of mobile apps and better support for cloud-oriented services, the term mobile cloud computing is introduced. Mobile cloud computing is basically an integration of cloud computing in the mobile environment. With the advent of mobile cloud computing, mobile users are taking advantage of new type of services and which facilities them in fully utilizing cloud computing services.

Mobile cloud computing [14,15] has the potential to transform the large arena of the IT industry. This will help in making software and hardware-oriented services more accessible and attractive [1]. One of the primary objectives of cloud computing is to provide computing and storage services at low and reasonable costs. This happens by sharing many resources between different users. The actual provisioning of such services at a low process depends on how efficiently resources are utilized in the cloud. A typical mobile cloud computing infrastructure is illustrated in Figure 1.

Cloud vendors can offer special hardware and particular software techniques for the provisioning of reliable services at a high price. Later, these reliable services could be sold to users by signing terms under certain clause or service level agreement. Nowadays, the cloud computing industry is using the term "no single point of failure". But the single point of failure often occurs when a single cloud service provider is hosting all these solutions [16]. It is worth mentioning that a true vendor-agnostic solutions will not only an open source technology solution (software) but will be accepted only if it is being operated on a vendor neutral hardware (by using off-the shelf, bare-metal/SDN-enabled devices) etc.

On the other hand, software-defined clouds (SDCs) make use of SDNs in order to create a programmable and flexible network by separation of functions for control plane and data plane. The reason for choosing SDNs in data center resource management is their simplicity and control over data center infrastructure. The idea of vendor agnostics through SDNs in data centers is implemented by the Open Flow with the decomposition of traffic control authorization to different parts [17–19]. The controller element is a powerful manager of the network that is processing information related to flows. Open Flow switches include basic functions like receiving, forwarding or looking up in a data traffic table. By using OpenFlow, routing is not confined to a Media Access Control (MAC) address or IP address. This basically helps in the determination of paths with the parameters of high security, low packet loss or low delay and also helps in maintaining the fine-grained scrutiny policies for various applications.

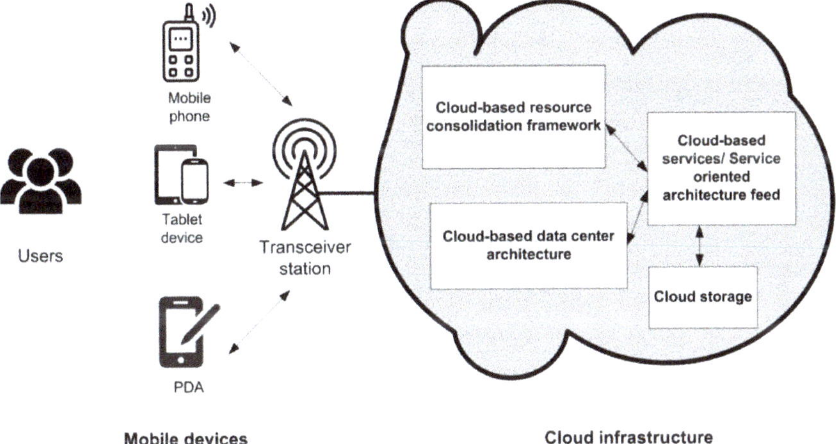

Figure 1. Mobile-oriented network infrastructure for cloud computing.

The main purpose of this paper is to contribute towards the field of SDCs by investigation of a major problem in cloud computing i.e. the placement of virtual machine (VM). VM placement is very important in datacenters and has been studied extensively, particularly for their use in the software-defined domain [20–22]. In a cloud environment, VMs are a key player as they provide the flexibility claimed by cloud service providers. Figure 1 presents the layout of VM in a distributed environment. When a computing service admitted into the cloud system demands for higher computational resources, then VM management plays a very important role. VM management helps in balancing the system constraints and loads [23,24]. Its main purpose is to retain user service satisfaction level. There are numerous VM placement challenges. Traditionally, the techniques for VM placement only focus on the resource allocation efficiency. Network research related to cloud resource management often focus on placement of VMs in data center environments. A vast number of VM placement techniques propose a solution based on available network resources [25]. This paper presents a relatively simpler approach for VM placement in the SDC environment. The concepts presented in this paper are related to the state-of-the-art technologies such as server and network resource utilization, software-defined networks, VM placement/ mapping and software-defined middle box networking. The paper presents a combination of these technologies for resource management in cloud environments.

The rest of the paper is further organized as follows. Section 2 discusses the related work, Section 3 presents the research allocation and mapping discussion in cloud environments, Section IV presents the mathematical modeling. In Section 4 we perform the performance evaluation. Finally, Section 5 concludes the paper.

2. Related Work

In cloud environments, resource sharing must be done in a way that a user's application requirements must not influence other user applications. Resource sharing must be done in such a way that these are secured and privately available [26]. VMs are acquired by applications on cloud infrastructure when needed. However, for cloud tenants, VM acquisition is a challenge. It is due to the limitations in cloud system's granularity and limitations in VM control and placement.

Data-intensive applications [27,28] frequently communicate with data centers. That is why the data traffic transmission of these applications is quite large. This results in network performance degradation and higher system overheads. VM placement strategies often use VM consolidation and reallocation techniques to solve vendor lock-in issues. These issues greatly influence network performance. In vendor lock-in issues, the users' traffic volumes can face delays. This ultimately leads to VM placement issues. The VM placement problem with traffic awareness [5,6] was proposed for solving these problems through network optimization based strategies.

In cloud environments, VMs follow certain patterns in accessing network resources. Research studies conducted in [29,30] involves a large number of CPU traces from different servers. It demonstrates that the demand traces are mostly in correlation and follow a periodic behavior. However, the concept of statistical multiplexing exists due to varying workloads. Data packet behavior for these applications relies primarily on the idea of exploiting possible correlations in VMs. Other approaches to vendor lock in issues include continuous monitoring of all VMs running on the network by using various VM measurement heuristics [31].

Current VM placement strategies have been extended for the inclusion of other data center infrastructure aspects such as network storage and network traffic. In a cloud infrastructure, all deployed VMs typically show a dependency on network traffic. The best optimization strategy to address their consolidation challenges is by hosting them on the nearest available physical machine [32,33]. Interestingly, network topology and data center design has a major impact on the selection of placement for traffic optimization targets [34]. Similar dependencies often occur for VMs and storage resources with different user requirements. In this situation, applications needing greater I/O performance can be moved closer to the storage locality.

Different vendors provide tools for resource-management functions. These tools include a wide range of applications. This includes system-level monitoring tools to application-level deep packet tracers and monitors. These sophisticated tools are a good choice; however, they slow down the system performance. Therefore, a vast-scale adoption of these tools will not only burden the network features, but will also influence underlying network resources (including virtual and infrastructure resources). In view of the above, a vendor-agnostic approach is used in [35,36] which proposes VM placement on a physical machine with the least data transfer time with respect to network bandwidth usage. However, within the datacenter premises, the data transmission rate is better due to wired communication. Therefore, users of these services expect high-qualityenterprise-level services rather than services offered by mobile devices with limited resources. Although not having enough tenant support for the VM migration, the cloud services provider have high control functions over all VMs locations. The manipulation of VMs can be performed by scaling in and out of physical resources.

Network support for tenant-controlledVM placement is difficult. An API-based SDN-enabled solution for these issues helps in providing a clean interface to the network administrator and is widely used in SDC environments.

SDNs [37,38] provide new possibilities for designing, operating, and securing data-intensive networks. However, the realization of these benefits largely requires the support of underlying infrastructure. In addition to handling the increased traffic loads, the network performance satisfaction opens new avenues of network services.

Mobile cloud computing based systems perform cloud computing functions with the exception that its users are mobile. Graph theory is a widely used concept in applied mathematics to structure pair wise models and relationships between objects. In this paper, we use graph-theoretic approach for resource consolidation on a vendor-agnostic hardware infrastructure which uses SDNs to administer network functions.

The proposed methodology is described by the formulation of a solution for VM placement that can be incorporated in SDCs. Currently, there is very limited support for VM placement in SDCs. For example, Amazon EC2 lacks support for co-locating its instance types. Although limited support features are present for cluster-based computational structures, high-performance features can only be afforded with premium prices [39,40]. Typically, the network resources available in close proximity are used for improved networking performance. It is believed that SDCs will be extensively used in future for flexibility and support in network applications and resource management.

3. Research Allocation and Mapping in Cloud Environments

Current mobility management schemes are based on centralized data access methodology. It is similar to the concepts used in traditional DC architectures. The main problem with this scheme is that it is difficult to manage. In terms of performance-based measures, these techniques results in routing and path optimization-related constraints which ultimately leads to performance degradation challenges.

The term vendor-agnostic refers to a concept where the products of a specific manufacturer are not tied to a particular vendor/brand etc. In distributed networks and systems theory, this term is often mingled with any off the self-solution. Vendor-agnostic solutions operate upon free, open-ended and generic solutions which involve basic mathematical optimization laws and principles not tied or related to a particular company. These solutions provide a clean interface to users for interacting with real-world problems. Our reason for highlighting vendor-agnostic behavior is based on the reason that we use a combination of open-ended hardware and software (via SDNs, graph-theory and Pareto-optimality) to achieve resource management functions.

In cloud environments, the rapid interaction between network's I/O devices, data and application services affect the system's overall performance [41]. The SDN concept to decouple data from control streams eases application and network performance. Here, we want to mention that SDN itself is an enabling technology. We need to employ SDN infrastructure for developing VM placement mechanism to achieve the desired goals. Therefore, we present a VM placement scheme for a SDC environment which can improve the service response time of applications.

We consider a software-defined cloud architecture where SDN based APIs administer the cloud resource management functions. These APIs manage topology and admission control features of cloud resources. Our framework's architecture is presented in Figure 2. Beneath the APIs lies a set of network manager and cloud manager. They control various functions of cloud e.g., mapping VMs, network statistics monitoring and controlling incoming outgoing packet requests. The last layer of the design architecture consists of virtual and physical resources.

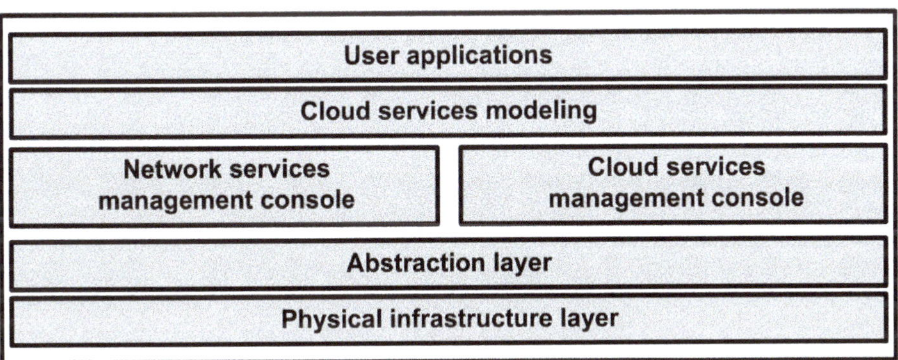

Figure 2. Proposed SDN cloud scenario.

SDN management APIs provide cloud resource management functions as high-level policies for the underlying network infrastructure. Such APIs help in managing and accessing an apparently infinite pool of computing resources like VMs etc. The function of the planner is to determine the location of hosting features for the received application requests in collaboration with cloud manager, modeler, and network manager. The modeler performs the comparison of received data and services from cloud planner and cloud manager. It is also used to model resource utilization features for updating network directory status. The network and cloud managers are used for managing virtual machines. The cloud and network managers on the other hand consolidate data at both physical and logical levels. The abstraction layer consists of logically-deployed physical hardware. Finally, the physical infrastructure layer consists of a list of physical resources that could be abstracted such as storage and network resources (routers or switches), servers, computing hosts, etc.

After sending off a request, the console of SDN manager makes sure that the made request is in compliance with the minimum number of SLAs. It then creates the topology of a blueprint. The topology information is later submitted to the admission controller. The admission controller validates and ensures that a connection can be established if current resources are sufficient for the proposed connection [42,43]. A simplistic approach of the performed sequential operation is illustrated in Figure 3.

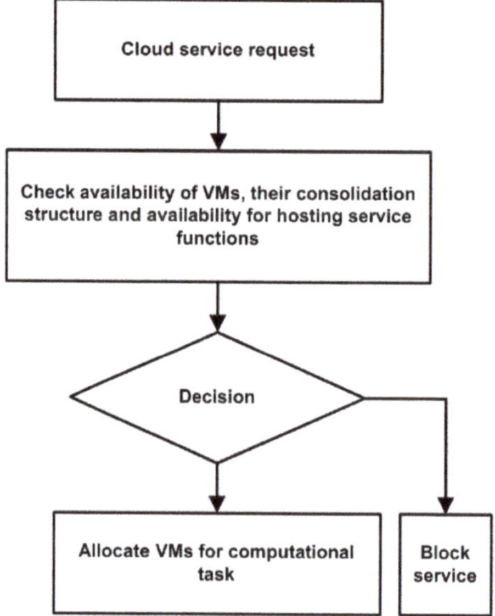

Figure 3. Virtual machine (VM) placement policy prototype in data center networks.

The location of hosting applications is determined by planner and modeler in consultation with cloud and network managers. Mapping of cloud resources is performed by the mapper. The proposed system performs VM placement. For ease of management, VM mapping should be controlled separately to ensure that cloud resources are managed in a clear and concise manner. The lower layer of SDC consists of different network resources. The layer for physical infrastructure contains any physical resource that could be abstracted e.g., storage and network resources (routers or switches), computing hosts, servers, etc. The abstraction layer provides abstraction information from a logical perspective. Conceptually physical layer resides beneath the abstraction layer [44,45].

In the proposed framework, by using graph theory, compute nodes are managed for allocation of VMs. In the proposed framework both virtual topologies and the physical infrastructure (switches, hosts, and links between them) are simulated for achieving dynamic routing features. In the presented scenario, all traffic patterns are supported by all the network elements. The assumptions in the proposed mechanism are mapped in a simulated environment for evaluation purpose.

The representation of the placement of VMs problem with the use of metrics (from linear algebra) is based on the fact that the cloud systems can be presented as a graph containing nodes and edges. Graph theory is the basic topic under discrete mathematics. Many advantages are present in the graph-theoretic approach.

On the basis of applied graph theory, we then manage compute nodes for VM allocation. In the proposed framework, both physical infrastructure and virtual topologies are simulated in CloudSim [46]. A detailed overview on latest trends and developments in the field of virtual resource management and network functions has been presented in [47].

4. Mathematical Modeling

We provide a mathematical representation of VM placement similar to [48] through our cloud model using a graph-theoretic approach. The entire interconnection between various entities of the proposed cloud is represented by adjacency matrices. Storage nodes (SN) represents data storage nodes. Compute nodes (CN) consist of multiple physical computational nodes, whereas data packet (DP) represents the data to be transmitted across the cloud. Our cloud infrastructure consists of 3 CNs, 2 SNs, and 3 DPs. Below we describe our model in detail. We consider a cloud system composed of m > 0 compute nodes (CN) and n > 0 storage nodes (SN). Please take note that the values of m and n are positive integers. The entire interconnection of the CNs and SNs can be depicted as a graph as shown in Figure 4.

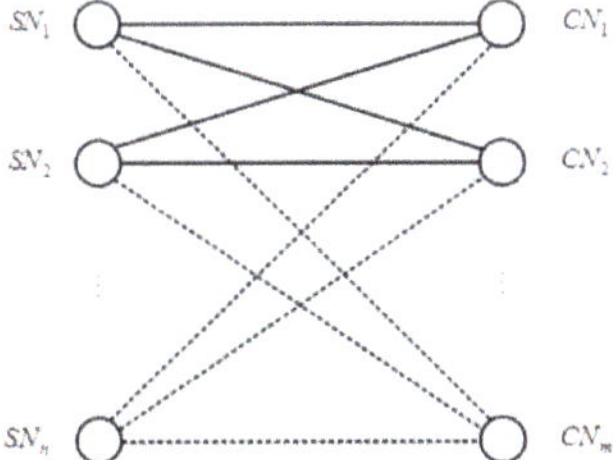

Figure 4. A graph-theoretic representation of the interconnection between the compute nodes (CN) and storage nodes (SN) forming a bipartite.

In discrete mathematics terminology (especially in graph theory), the graph shown in Figure 5 is known as a bipartite. A bipartite is a group of two sets of nodes where each member of each set is able to "communicate" with each and every member of the other set. The edges connecting the CN and SN may represent any relationship between these nodes. In order to limit and scale down the performance of our simulation, we assume that these edges could represent either bandwidth in MBps or time constant in secs/MB (which is just the reciprocal of the bandwidth). For example, the edge connecting SN1 to CN1 could represent the bandwidth value of 3.2 MBps or time constant of 0.3125 secs/MB (i.e., 1/3.2 MBps).

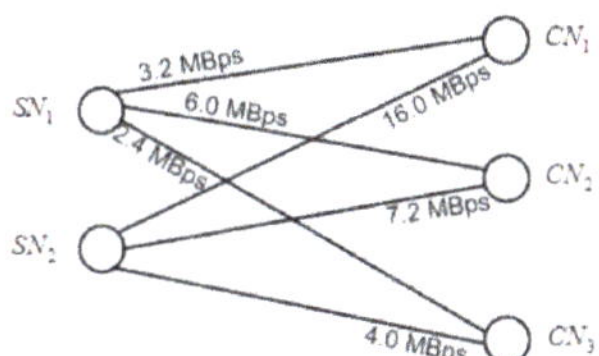

Figure 5. A graph-theoretic representation of a 2-SN, 3-CN cloud system given the values of the networks bandwidths between each combination of CN and SN nodes. This is an example of a 2×3 bipartite B3.2.

In applied graph theory, an adjacency matrix is a matrix that represents the values of all edges connected to the nodes in the graph. Consider the $n \times m$ adjacency matrix

$$B = \left[b_{ij} \right]_{n \times m} = \begin{bmatrix} b_{11} & b_{12} & \cdots & b_{1m} \\ b_{21} & b_{22} & \cdots & b_{2m} \\ \vdots & \vdots & \ddots & \vdots \\ b_{n1} & b_{n2} & \cdots & b_{nm} \end{bmatrix}, \tag{1}$$

where the matrix elements b_{ij} for $1 \leq i \leq n$ and $1 \leq j \leq m$ are values representing the edges connecting node i to node j. As we consider a 2-CN, 3-SN cloud system as depicted in Figure 6, the network bandwidth between the nodes are as follows:

- SN1—CN1: 3.2 MBps;
- SN1—CN2: 6.0 MBps;
- SN1—CN3: 2.4 MBps;
- SN2—CN1: 16.0 MBps;
- SN2—CN2: 7.2 MBps; and
- SN2—CN3: 4.0 MBps;

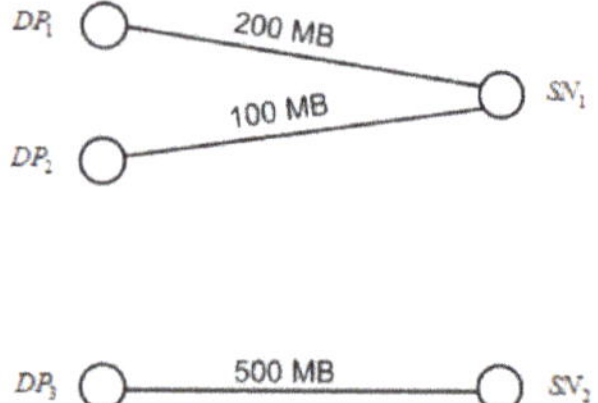

Figure 6. A graph representation of the relationship between the data pieces and the storage nodes. The first sub-graph is a 2 × 1 bipartite while the other sub-graph is a 1 × 1 bipartite (or simply a connection between two nodes).

Then the corresponding graph-theoretic representation will result in Figure 5 having the adjacency matrix with row i representing the SN number and column j representing the CN number.

$$\mathbf{B}_{3.2} \begin{bmatrix} 3.2 & 16.0 \\ 6.0 & 7.2 \\ 2.4 & 4.0 \end{bmatrix} \text{MBps.} \tag{2}$$

Similar to the approach in the previous section, a graph-theoretic approach can also be used to represent the relationship between the data pieces and SNs. In order to consider an environment two data pieces, DP1 = 200 MB and DP2 = 100 MB are both stored at storage node SN1, while another data piece, DP3 = 500 MB, is stored in SN2. The resulting graph shall be composed of two sub-graphs: one graph representing the relationship between DP1 and DP2 to SN1, and the relationship between DP3 to SN2. Please take note that each sub-graph is also a bipartite as shown in Figure 6. Since each data piece is stored only in a dedicated SN, it will be assumed in this architecture that the data piece is not shared between other SNs. Therefore, each sub-graph will only have one SN but can have multiple DPs.

The entire interconnection between the DP and SN can also be represented by an adjacency matrix where the column q shall represent the SN number. Since there are 2 SNs, then the matrix will have $n = 2$ columns. The number of rows of the adjacency matrix shall be equal to the maximum number of data pieces in any SN. In this particular example, since SN1 has two data pieces, namely DP1 and DP2, the number of rows shall be equal to $p = 2$. The resulting adjacency matrix becomes

$$\mathbf{D}_{2.2} = [d_{ki}]_{2\times2} = \begin{bmatrix} DP_1 & DP_3 \\ DP_2 & 0 \end{bmatrix} = \begin{bmatrix} 200 & 500 \\ 100 & 0 \end{bmatrix} MB. \tag{3}$$

Consider for example a network bandwidth of b MBps. If data of size d MB will be transmitted into the network, then the response time can be obtained through

$$t = \frac{d}{b}[sec], \tag{4}$$

where simply the data and bandwidth are being adjusted with respect to time.

Similarly, if the network bandwidth b is inverted resulting in the time constant τ = 1/b (secs./MB), then the response time can be calculated using

$$t = d(MB).\,\tau\left(\frac{sec}{MB}\right)[sec], \tag{5}$$

where time and bandwidth are relating to resource requirements

However, this expression is only valid for scalar quantities, i.e., if there is only one data piece being processed by one CN through one SN. In fact, the notation "×" can be used here to represent scalar multiplication.

First, consider the graphical representation of the merger between $B_{n,m}$ and $D_{p,n}$ as shown in Figure 7. From the graphical representation, data flows between the SN and CN and DPs storage location can be seen. Using graph theory [16], it is possible to graphically represent networks using nodes and edges even if their quantities are different. The matrix can be defined as

$$\Upsilon_{n\times m} = \left[\tau_{ij}\right]_{n\times m} \tag{6}$$

where the values τ_{ij} represent the time constants between the SN and CN. Let us call this the time constant matrix. Basically, the values in this matrix are just the reciprocals of the bandwidths, therefore the following mathematical expression

$$\left[\tau_{ij}\right]_{n\times m} = \left[\frac{1}{b_{ij}}\right]_{n\times m} \tag{7}$$

where cross matrix multiplication ensures SN to CN matrix mapping shall apply for $1 \leq i \leq n$ and $1 \leq j \leq m$. Given the data set matrix $\mathbf{D}p,n$ we can now get the response times for each data piece in various CNs.

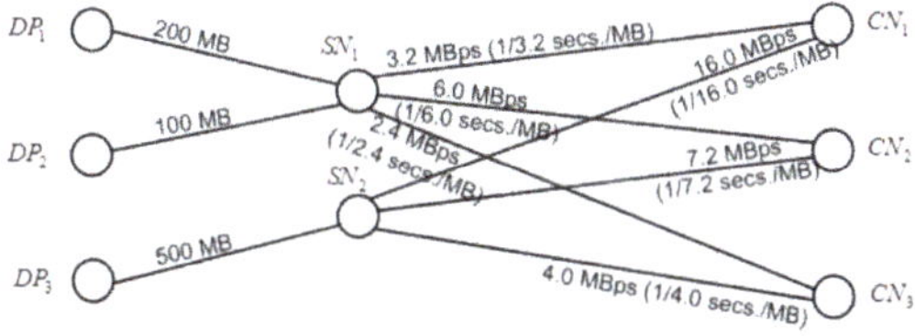

Figure 7. Graph representing entire cloud mapping.

The series of manual computations above are easily done due to the small dimensions of the cloud system. Since real cloud systems have hundreds of thousands of SN and CN, it will be impossible for us to have all the combinations and compute them manually. It is now important to have everything done with a computer through linear algebra. In order to obtain the total response times tCN,1, tCN,2 and tCN,3 we need the following step:

(1) Step 1: get the time constant matrix from the bandwidth matrix: Let

$$B_{3.2} = \begin{bmatrix} 3.2 & 16.0 \\ 6.0 & 7.2 \\ 2.4 & 4.0 \end{bmatrix} \text{MBps} \tag{8}$$

be the bandwidth matrix of the cloud system. The time constants for each element in the matrix can be obtained by simply getting the reciprocals of each element. The resulting matrix becomes

$$B_{3.2} = \begin{bmatrix} 1/3.2 & 1/16.0 \\ 1/6.0 & 1/7.2 \\ 1/2.4 & 1/4.0 \end{bmatrix} = \begin{bmatrix} 0.3125 & 0.6250 \\ 0.1667 & 0.1389 \\ 0.4167 & 0.2500 \end{bmatrix} \text{sec /MB.} \tag{9}$$

(2) Step 2: get the transpose of the time constant matrix: Given the time constant matrix $\Upsilon_{3.2}$, its transpose can be obtained as

$$\Upsilon_{3.2}^{T} = \begin{bmatrix} 0.3125 & 0.6250 \\ 0.1667 & 0.1389 \\ 0.4167 & 0.2500 \end{bmatrix} = \begin{bmatrix} 0.3125 & 0.1667 & 0.4167 \\ 0.6250 & 0.1389 & 0.2500 \end{bmatrix} \text{sec /MB}^{T}. \tag{10}$$

(3) Step 3: multiply the data set matrix with the transposed time constant matrix: Given the data set matrix

$$\mathbf{D}_{2.2} = \begin{bmatrix} 200 & 500 \\ 100 & 0 \end{bmatrix} \text{MB.} \tag{11}$$

The response time matrix can then be obtained as follows

$$T_{R} = \mathbf{D}_{2.2}\Upsilon_{3.2}^{T} = \begin{bmatrix} DP_1 & DP_3 \\ DP_2 & 0 \end{bmatrix} \begin{bmatrix} \tau_{11} & \tau_{12} \\ \tau_{21} & \tau_{22} \\ \tau_{31} & \tau_{32} \end{bmatrix} = \begin{bmatrix} DP_1 & DP_3 \\ DP_2 & 0 \end{bmatrix} \begin{bmatrix} \tau_{11} & \tau_{21} & \tau_{31} \\ \tau_{12} & \tau_{22} & \tau_{32} \end{bmatrix}^{T} \tag{12}$$

$$T_{R} = \begin{bmatrix} 200 & 500 \\ 100 & 0 \end{bmatrix} \begin{bmatrix} 0.3125 & 0.1667 & 0.4167 \\ 0.6250 & 0.1389 & 0.2500 \end{bmatrix} \tag{13}$$

$$T_{R} = \begin{bmatrix} (t_{11} + t_{31}) & (t_{12} + t_{32}) & (t_{13} + t_{33}) \\ (t_{21}) & (t_{22}) & (t_{23}) \end{bmatrix} = \begin{bmatrix} t_{R,11} & t_{R,12} & t_{R,13} \\ t_{R,21} & t_{R,22} & t_{R,23} \end{bmatrix}. \tag{14}$$

One key characteristic of the network response time matrix T_R is that if you get the sum of all elements per column, you actually obtain the total response time for each and every CN. Each column of T_R represents each CN. Since, in this example, there are 3 CNs in the cloud system, T_R results in a matrix having 3 columns as well. If the matrix is denoted by

$$\mathbf{T_R} = \begin{bmatrix} t_{R,11} & t_{R,12} & t_{R,13} \\ t_{R,21} & t_{R,22} & t_{R,23} \end{bmatrix}, \tag{15}$$

The total response time per CN can be obtained using the expression

$$t_{CN,i} = \sum_{j=1}^{n} t_{R,ji}, \tag{16}$$

Therefore,

$$t_{CN,1} = \sum_{j=1}^{n=2} t_{R,j1} = t_{R,11} + t_{R,21} \tag{17}$$

$$t_{CN,2} = \sum_{j=1}^{n=2} t_{R,j2} = t_{R,12} + t_{R,22} \tag{18}$$

$$t_{CN,3} = \sum_{j=1}^{n=2} t_{R,j3} = t_{R,13} + t_{R,23} \tag{19}$$

Real-world optimization issues can be implemented on cloud-based systems to involve multiple conflicting objectives. Therefore, a vector-optimization problem in a standardized manner can be represented as a standardized vector $X = (x_1, x_1, \ldots, x_n)$. A Pareto-optimal solution [49,50] for resource existence (when no other solution exists) is represented in Figure 8. This helps in ensuring that one objective (resource allocation) can be improved without affecting the other objective.

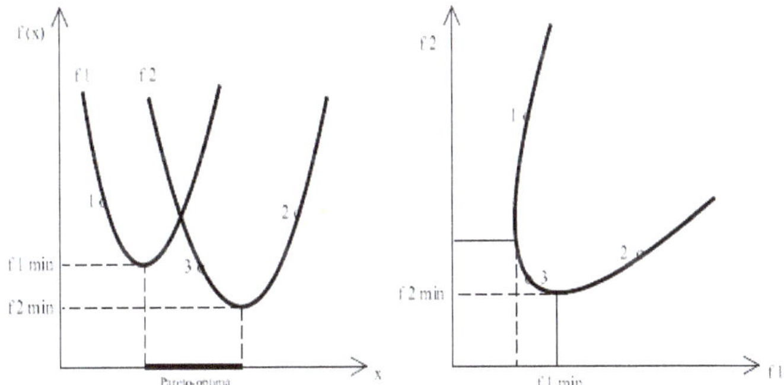

Figure 8. The Pareto-optimal solution where one objective can be improved without the expense of others.

5. Performance Evaluation

In this section, we explain the simulations and experiments carried to evaluate the proposed resource consolidation approach. We used a channel model approach similar to presented in [51,52], which is widely used for mobile cloud and cellular networks. The testbed simulation consisted of two storage nodes, three data sets, and three mobile hosts. The mobile devices map their resources by using a graph-theoretic model as explained in previous sections and use SDN-based infrastructure for controlling data and traffic behavior functions. In this regard, we implemented our algorithm on CloudSim v 3.0. The Cloudsim is often used as an extensible simulation toolkit for simulation purposes.

Algorithm 1 VM Placement

1:	CN denotes the Computes of cloud
2:	$t_{CN,i}$ denotes the response time value of CNi
3:	*least* denotes least response time value of CNs
4:	j denotes CN having least response time value
5:	For calculating t_R for each CN, we have
6:	$t_{CN,i} = \sum_{j=1}^{n} t_{R,\,ji}$
7:	$i \leftarrow 0$
8:	$j \leftarrow 0$
9:	$least \leftarrow t_{CN,0}$
10:	**while** $i < n$ **do**
11:	**if** *least* $> t_{CN,i}$:**then**
12:	$least \leftarrow t_{CN,i}$
13:	$j \leftarrow i$
14:	**end if**
15:	$i \leftarrow i+1$
16:	**end while**
17:	Choose VM location at CN j
18:	**Exit**

We developed an algorithm (Algorithm 1) for virtual machine placement on a particular cloud node. It works by calculating service response time of individual compute nodes and then selecting VM with least response time. By using the proposed Algorithm 1, we compute the service response time T_R of individual CN and select a CN having least response time to host VM. We then calculated the response time of these data loads using vmallocationpolicysimple algorithm [53,54]. We selected the vmallocationpolicysimple algorithm because of two major reasons. Firstly, it does not implement dynamic consolidation of VMs, and only places new VMs on hosts; fulfilling our scenario's demand. Secondly, it is the default VM placement strategy in CloudSim. Finally, we compared the service response time of algorithm 1 with that of vmallocationpolicysimple algorithm for given workloads. The evaluation setting is similar to emulate the environment presented in Figure 7.

Simulation results in Figure 9 show the service response time for tasks requesting DP1 data load. In Figures 10 and 11, we illustrate the service response time for tasks requesting DP2 and DP3 respectively. The graphics illustrations reveal that our proposed scheme demonstrates improved service response time to requests as compared to vmallocationpolicysimple algorithm. It is because the presented algorithm clearly chooses a CN with reduced response time and shorter data access route for VM allocation.

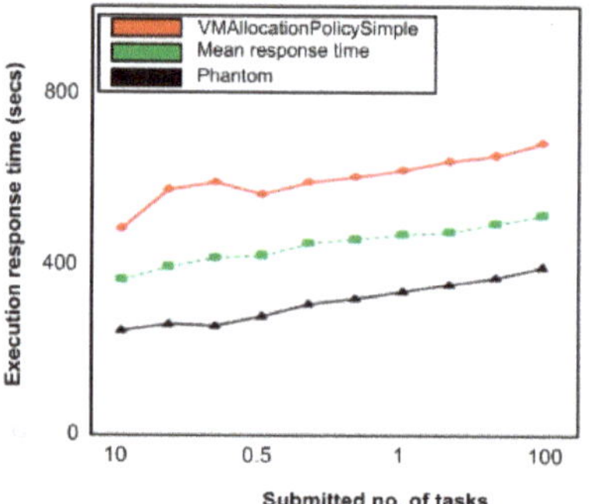

Figure 9. Service response time (DP1 requesting tasks).

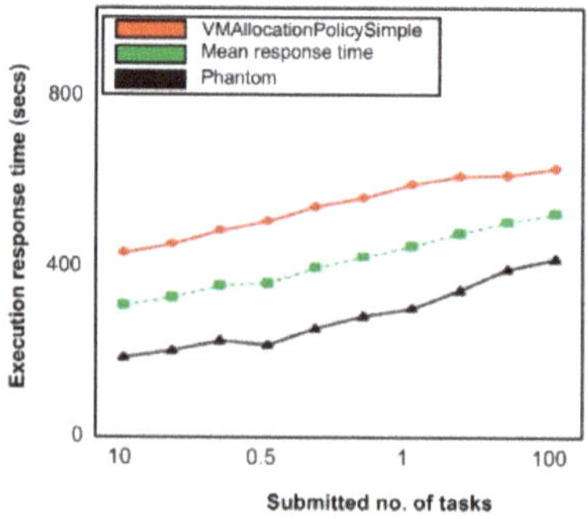

Figure 10. Service response time (DP2 requesting tasks).

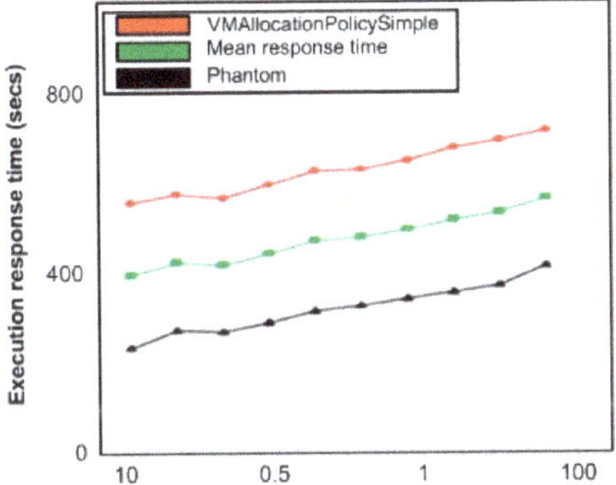

Figure 11. Service response (DP3 requesting tasks).

By observing the comparative results in Figure 12, it can be observed that the traffic intensity for DP1, DP2, and DP3 exhibit the same behavior of variance with respect to mean response time. However, the difference in values with respect to the vmallocationpolicysimple is different. This also resulted in increased variance rates of bandwidth consumption. In order to retain the job queue waiting time, we can manage the waiting-time window slot timing. If the received task arrives which can be completed in a relatively long period of time, the mapping scheme can adjust itself according to window time and accommodates more tasks as compared to the vmallocationpolicysimple strategy. The same concept can be improved with the predictive analysis however we didn't consider it due to overhead costs incurred on VM loads.

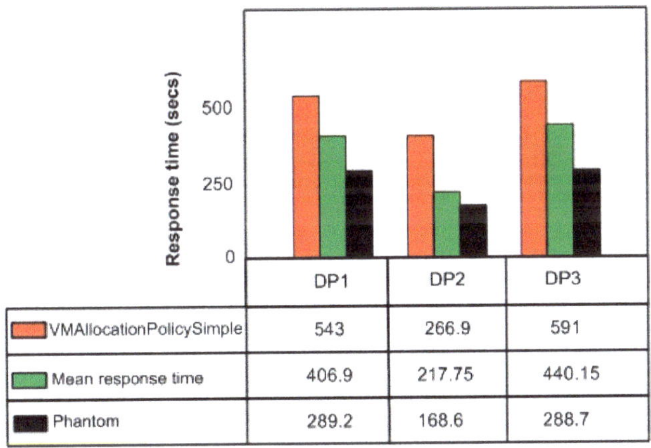

	DP1	DP2	DP3
VMAllocationPolicySimple	543	266.9	591
Mean response time	406.9	217.75	440.15
Phantom	289.2	168.6	288.7

Figure 12. Cumulative comparison of service response time for 100 requests.

6. Conclusions and Future Work

Future mobile-oriented networks are taking the computing industry by storm. The recent developments witnessed in the enhanced computational capability of mobile equipment led to the concepts of mobile cloud computing. Taking this paradigm to another step, in this paper we presented a case where cloud data centers are managed in a mobile cloud environment. We begin the paper by explaining the importance of mobile future network architectures followed by the concepts of resource management in mobile clouds using a vendor-agnostic approach (through SDNs).

To sum up the concept of the paper, we argue that the cloud computing concept involves the availability of computing resources for data storage and processing. Due to increasing number of network applications, number of users and their requirements, there is a dire need to develop tools to improve cloud computing performance. On the other hand, software-defined networking concept allows cloud data center administers to manage cloud resource allocation function according to their own needs via bypassing proprietary network peripherals. As SDN concepts discourage excessive use of proprietary equipment, it is often referred as a bare-metal solution, off-the-shelf solution and vendor-agnostic solution).

The paper relates SDN-based mobile cloud environment to propose a VM mapping policy using a graph-theoretic approach. The reason for calling this technique vendor-agnostic is the use of a vendor-agnostic platform (SDN-enabled H/W) for evaluation purpose.

In this paper, the simple representation of VM resource allocation and representation helped in clearly determine the network management by the use of matrices. A vendor-agnostic-based approach, therefore, offers several advantages over conventional approaches. These advantages can be seen particularly in distributed systems like cloud computing environments. Therefore, we implemented this approach to consolidate VM resources in a simplistic and well-organized way. We believe that graphs can be represented using adjacency matrices where each element of the matrix denotes values that show relationships between any two nodes. Therefore, we used a graph-theoretic approach to achieve our consolidation approach. We developed a framework and compared its performance with the vmallocationpolicysimple technique. Results demonstrate that our proposed framework can limit the cloud topology scaling issues of VM placement in a more clearer and concise manner. We strongly believe that a vendor-agnostic approach in data centers can be considered as the next step towards the evolution of virtualization, mobile cloud computing, and future mobile-oriented networks.

Author Contributions: Conceptualization, A.A.A.; data curation, A.A.A. and M.A.A.-q.; formal analysis, M.A.A.-q. and A.H.; investigation, A.A.A. and A.H.; methodology, M.A.E. and A.A.E.; resources, M.A.E. and S.J.; software, A.A.E.; supervision, S.K.; writing—original draft, A.A.A.; writing—review and editing, A.A.A. and S.J. All authors have read and approved the manuscript.

Funding: Following are results of a study on the "Leaders in INdustry-university Cooperation+" project, supported by the Ministry of Education and National Research Foundation of Korea.

Acknowledgments: The authors thank the anonymous reviewers for their thorough feedback and suggestions, which were crucial to improving the contents and presentation of the article.

Conflicts of Interest: The authors declare no conflict of interest.

References

1. Kim, J.I.; Choi, N.J.; You, T.W.; Jung, H.; Kwon, Y.W.; Koh, S.J. Mobile-Oriented Future Internet: Implementation and Experimentations over EU–Korea Testbed. *Electronics* **2019**, *8*, 338. [CrossRef]
2. Kim, J.I.; Jung, H.; Koh, S.J. Mobile oriented future internet (MOFI): Architectural design and implementations. *ETRI J.* **2013**, *35*, 666–676. [CrossRef]
3. Jung, H.Y.; Koh, S.J. *Mobile-Oriented Future Internet (MOFI): Architecture and Protocols*; ETRI: Daejeon, Korea, 2010.
4. Boukerche, A.; Guan, S.; Grande, R.E.D. Sustainable Offloading in Mobile Cloud Computing: Algorithmic Design and Implementation. *ACM Comput. Surv.* **2019**, *52*, 11. [CrossRef]
5. Chaudhry, S.A.; Kim, I.L.; Rho, S.; Farash, M.S.; Shon, T. An improved anonymous authentication scheme for distributed mobile cloud computing services. *Clust. Comput.* **2019**, *22*, 1595–1609. [CrossRef]
6. Ahmed, A.A.; Alzahrani, A.A. A comprehensive survey on handover management for vehicular ad hoc network based on 5G mobile networks technology. *Trans. Emerg. Telecommun. Technol.* **2019**, *30*, e3546. [CrossRef]
7. Kushwah, R.; Tapaswi, S.; Kumar, A. A detailed study on Internet connectivity schemes for mobile ad hoc network. *Wirel. Pers. Commun.* **2019**, *104*, 1433–1471. [CrossRef]
8. Garcia, A.J.; Toril, M.; Oliver, P.; Luna-Ramirez, S.; Garcia, R. Big Data Analytics for Automated QoE Management in Mobile Networks. *IEEE Commun. Mag.* **2019**, *57*, 91–97. [CrossRef]
9. Liu, K.; Zha, Z.; Wan, W.; Aggarwal, V.; Fu, B.; Chen, M. Optimizing TCP Loss Recovery Performance Over Mobile Data Networks. *IEEE Trans. Mob. Comput.* **2019**. [CrossRef]
10. Aceto, G.; Ciuonzo, D.; Montieri, A.; Pescapé, A. Mobile encrypted traffic classification using deep learning: Experimental evaluation, lessons learned, and challenges. *IEEE Trans. Netw. Serv. Manag.* **2019**. [CrossRef]

11. Misra, S.; Wolfinger, B.E.; Achuthananda, M.P.; Chakraborty, T.; Das, S.N.; Das, S. Auction-Based Optimal Task Offloading in Mobile Cloud Computing. *IEEE Syst. J.* **2019**. [CrossRef]

12. Moreno-Vozmediano, R.; Huedo, E.; Montero, R.S.; Llorente, I.M. A Disaggregated Cloud Architecture for Edge Computing. *IEEE Internet Comput.* **2019**, *23*, 31–36. [CrossRef]

13. Chen, D.; Zhang, X.; Wang, L.L.; Han, Z. Prediction of Cloud Resources Demand Based on Hierarchical Pythagorean Fuzzy Deep Neural Network. *IEEE Trans. Serv. Comput.* **2019**. [CrossRef]

14. Ahmed, E.; Naveed, A.; Gani, A.; Ab Hamid, S.H.; Imran, M.; Guizani, M. Process state synchronization-based application execution management for mobile edge/cloud computing. *Future Gener. Comput. Syst.* **2019**, *91*, 579–589. [CrossRef]

15. Agrawal, N.; Shashikala, T. A trustworthy agent-based encrypted access control method for mobile cloud computing environment. *Perv. Mob. Comput.* **2019**, *52*, 13–28. [CrossRef]

16. Sharma, Y.; Weisheng, S.; Daniel, S.; Bahman, J. Failure-aware energy-efficient VM consolidation in cloud computing systems. *Future Gener. Comput. Syst.* **2019**, *94*, 620–633. [CrossRef]

17. Minh, Q.T.; Dang, T.K.; Nam, T.; Kitahara, T. Flow aggregation for SDN-based delay-insensitive traffic control in mobile core networks. *IET Commun.* **2019**, *13*, 1051–1060. [CrossRef]

18. Wu, W.; Liu, J.; Huang, T. Decoupled delay and bandwidth centralized queue-based QoS scheme in OpenFlow networks. *China Commun.* **2019**, *16*, 70–82.

19. Shamshirband, S.; Hossein, S. LAAPS: An efficient file-based search in unstructured peer-to-peer networks using reinforcement algorithm. *Int. J. Comput. Appl.* **2018**. [CrossRef]

20. Abbasi, A.A.; Al-qaness, M.A.; Elaziz, M.A.; Khalil, H.A.; Kim, S. Bouncer: A Resource-Aware Admission Control Scheme for Cloud Services. *Electronics* **2019**, *8*, 928. [CrossRef]

21. Abbasi, A.A.; Abbasi, A.; Shamshirband, S.; Chronopoulos, A.T.; Persico, V.; Pescapè, A. Software-Defined Cloud Computing: A Systematic Review on Latest Trends and Developments. *IEEE Access* **2019**, *7*, 93294–93314. [CrossRef]

22. Jin, H.; Abbasi, A.A.; Wu, S. Pathfinder: Application-aware distributed path computation in clouds. *Int. J. Parallel Program.* **2017**, *45*, 1273–1284. [CrossRef]

23. Priya, B.; Gnanasekaran, T. To optimize load of hybrid P2P cloud data-center using efficient load optimization and resource minimization algorithm. *Pee Peer Netw. Appl.* **2019**. [CrossRef]

24. Curino, C.; Subru, K.; Konstantinos, K.; Sriram, R.; Giovanni, M.F.; Botong, H.; Kishore, C.; Arun, S.; Chen, Y.; Heddaya, S.; et al. Hydra: A federated resource manager for data-center scale analytics. In Proceedings of the 16th USENIX Symposium on Networked Systems Design and Implementation (NSDI '19), Boston, MA, USA, 26–28 February 2019; NetApp: Sunnyvale, CA, USA, 2019; pp. 177–192.

25. Asghari, S.; Navimipour, N.J. Resource discovery in the peer to peer networks using an inverted ant colony optimization algorithm. *Peer Peer Netw. Appl.* **2019**, *12*, 129–142. [CrossRef]

26. Shamshirband, S.; Chronopoulos, A.T. A new malware detection system using a high performance-ELM method. In Proceedings of the 23rd International Database Applications & Engineering Symposium, ACM, Athens, Greece, 10–12 June 2019; p. 33.

27. Shuib, L.; Shamshirband, S.; Ismail, M.H. A review of mobile pervasive learning: Applications and issues. *Comput. Hum. Behav.* **2015**, *46*, 239–244. [CrossRef]

28. Stan, R.G.; Catalin, N.; Florin, P. Cloudwave: Content gathering network with flying clouds. *Future Gener. Comput. Syst.* **2019**, *98*, 474–486. [CrossRef]

29. Lin, L.; Liu, X.; Ma, R.; Li, J.; Wang, D.; Guan, H. vSimilar: A high-adaptive VM scheduler based on the CPU pool mechanism. *J. Syst. Archit.* **2019**. [CrossRef]

30. Kalogirou, C.; Koutsovasilis, P.; Antonopoulos, C.D.; Bellas, N.; Lalis, S.; Venugopal, S.; Pinto, C. Exploiting CPU Voltage Margins to Increase the Profit of Cloud Infrastructure Providers. In Proceedings of the 2019 19th IEEE/ACM International Symposium on Cluster, Cloud and Grid Computing (CCGRID), Larnaca, Cyprus, 14–17 May 2019; pp. 302–311.

31. Moges, F.F.; Abebe, S.L. Energy-aware VM placement algorithms for the OpenStack Neat consolidation framework. *J. Cloud Comput.* **2019**, *8*, 2. [CrossRef]

32. Le, F.; Nahum, E.M. Experiences Implementing Live VM Migration over the WAN with Multi-Path TCP. In Proceedings of the IEEE Infocom 2019 IEEE Conference on Computer Communications, Paris, France, 29 April–2 May 2019; pp. 1090–1098.

33. Mohiuddin, I.; Ahmad, A. Workload aware VM consolidation method in edge/cloud computing for IoT applications. *J. Parallel Distrib. Comput.* **2019**, *123*, 204–214. [CrossRef]
34. Zhao, H.; Han, G.; Niu, X. The Signal Control Optimization of Road Intersections with Slow Traffic Based on Improved PSO. *Mob. Netw. Appl.* **2019**. [CrossRef]
35. Guerrero, C.; Isaac, L.; Carlos, J. A lightweight decentralized service placement policy for performance optimization in fog computing. *J. Ambient Intell. Humaniz. Comput.* **2019**, *10*, 2435–2452. [CrossRef]
36. Badawy, M.; Hisham, K.; Hesham, A. New approach to enhancing the performance of cloud-based vision system of mobile robots. *Comput. Electr. Eng.* **2019**, *74*, 1–21. [CrossRef]
37. Lin, F.P.C.; Tsai, Z. Hierarchical Edge-Cloud SDN Controller System with Optimal Adaptive Resource Allocation for Load-Balancing. *IEEE Syst. J.* **2019**. [CrossRef]
38. Chirivella-Perez, E.; Marco-Alaez, R.; Hita, A.; Serrano, A.; Alcaraz Calero, J.M.; Wang, Q.; Neves, P.M.; Bernini, G.; Koutsopoulos, K.; Martínez Pérez, G.; et al. SELFNET 5G mobile edge computing infrastructure: Design and prototyping. *Softw. Pract. Exp.* **2019**. [CrossRef]
39. Ma, X.; Wang, S.; Zhang, S.; Yang, P.; Lin, C.; Shen, X.S. Cost-Efficient Resource Provisioning for Dynamic Requests in Cloud Assisted Mobile Edge Computing. *IEEE Trans. Cloud Comput.* **2019**. [CrossRef]
40. Tasiopoulos, A.; Ascigil, O.; Psaras, I.; Toumpis, S.; Pavlou, G. FogSpot: Spot Pricing for Application Provisioning in Edge/Fog Computing. *IEEE Trans. Serv. Comput.* **2019**. [CrossRef]
41. Rehman, A.; Hussain, S.S.; ur Rehman, Z.; Zia, S.; Shamshirband, S. Multi-objective approach of energy efficient workflow scheduling in cloud environments. *Concurr. Comput. Pract. Exp.* **2019**, *31*, e4949. [CrossRef]
42. Al-Dulaimi, A.; Mumtaz, S.; Al-Rubaye, S.; Zhang, S.; Chih-Lin, I. A Framework of Network Connectivity Management in Multi-Clouds Infrastructure. *IEEE Wirel. Commun.* **2019**. [CrossRef]
43. Park, S.T.; Oh, M.R. An empirical study on the influential factors affecting continuous usage of mobile cloud service. *Clust. Comput.* **2019**, *22*, 1873–1887. [CrossRef]
44. Zhou, Y.; Tian, L.; Liu, L.; Qi, Y. Fog computing enabled future mobile communication networks: A convergence of communication and computing. *IEEE Commun. Mag.* **2019**, *57*, 20–27. [CrossRef]
45. Elhabbash, A.; Faiza, S.; James, H.; Yehia, E. Cloud brokerage: A systematic survey. *ACM Comput. Surv.* **2019**, *51*, 119. [CrossRef]
46. Barbierato, E.; Gribaudo, M.; Iacono, M.; Jakóbik, A. Exploiting CloudSim in a multiformalism modeling approach for cloud based systems. *Simul. Model. Pract. Theory* **2019**, *93*, 133–147. [CrossRef]
47. Laghrissi, A.; Taleb, T. A survey on the placement of virtual resources and virtual network functions. *IEEE Commun. Surv. Tutor.* **2018**, *21*, 1409–1434. [CrossRef]
48. Piao, J.T.; Yan, J. A network-aware virtual machine placement and migration approach in cloud computing. In Proceedings of the 2010 Ninth International Conference on Grid and Cloud Computing, Nanjing, China, 1–5 November 2010; pp. 87–92.
49. Blasco, X.; Reynoso-Meza, G.; Sánchez-Pérez, E.A.; Sánchez-Pérez, J.V. Computing optimal distances to pareto sets of multi-objective optimization problems in asymmetric normed lattices. *Acta Appl. Math.* **2019**, *159*, 75–93. [CrossRef]
50. Wang, S.; Zhao, Y.; Xu, J.; Yuan, J.; Hsu, C.H. Edge server placement in mobile edge computing. *J. Parallel Distrib. Comput.* **2019**, *127*, 160–168. [CrossRef]
51. Hong, S.T.; Kim, H. QoE-aware Computation Offloading to Capture Energy-Latency-Pricing Tradeoff in Mobile Clouds. *IEEE Trans. Mob. Comput.* **2018**. [CrossRef]
52. Lei, L.; Xu, H.; Xiong, X.; Zheng, K.; Xiang, W. Joint Computation Offloading and Multi-User Scheduling using Approximate Dynamic Programming in NB-IoT Edge Computing System. *IEEE Internet Things J.* **2019**. [CrossRef]
53. CloudSim. Available online: http://www.cloudbus.org/cloudsim/doc/api/org/cloudbus/cloudsim/power/PowerVmAllocationPolicySimple.html (accessed on 7 October 2019).
54. Jammal, M.; Hawilo, H.; Kanso, A.; Shami, A. Generic input template for cloud simulators: A case study of CloudSim. *Softw. Pract. Exp.* **2019**, *49*, 720–747. [CrossRef]

Article

TARCS: A Topology Change Aware-Based Routing Protocol Choosing Scheme of FANETs

Jie Hong [1,2,*] **and Dehai Zhang** [1]

[1]   Key Laboratory of Microwave Remote Sensing, National Space Science Center, Chinese Academy of Sciences, Beijing 100190, China; zhangdehai@mirslab.cn

[2]   University of Chinese Academy of Sciences, Beijing 100094, China

*   Correspondence: hjane666@126.com; Tel.: +86-131-4603-6664

Received: 21 December 2018; Accepted: 21 February 2019; Published: 2 March 2019

Abstract: The rapid change of topology is one of the most important factors affecting the performance of the routing protocols of flying ad hoc networks (FANETs). A routing scheme suitable for highly dynamic mobile ad hoc networks is proposed for the rapid change of topology in complex scenarios. In the scheme moving nodes sense changes of the surrounding network topology periodically, and the current mobile scenario is confirmed according to the perceived result. Furthermore, a suitable routing protocol is selected for maintaining network performances at a high level. The concerned performance metrics are packet delivery ratio, network throughput, average end-to-end delay and average jitter. The experiments combine the random waypoint model, the reference point group mobility model and the pursue model to a chain scenario, and simulate the large changes of the network topology. Results show that an appropriate routing scheme can adapt to rapid changes in network topology and effectively improve network performance.

Keywords: flying ad hoc network (FANET); mobile ad hoc network (MANET); highly dynamic; periodical; topology change awareness; routing protocol

---

1. Introduction

Mobility is one of the most prominent features of mobile ad hoc networks (MANETs) and one of the important factors affecting network performance. Being a typical subset of MANETs, the node mobility in flying ad hoc networks (FANETs) [1–3] is stronger and severely impacts the network performance (Figure 1).

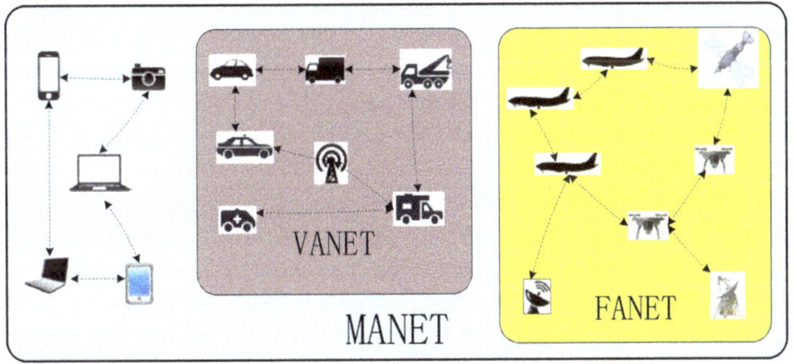

Figure 1. MANET, VANET and FANET.

FANET has some unique characteristics compared to MANET and VANET (Table 1). In FANETs, high-speed moving nodes usually need to complete multiple tasks, such as search/exploration, reconnaissance/patrol, or target tracking. Different task scenarios could have corresponding node mobility models [1]. When the task changes, the node mobility mode (such as node motion speed, direction, distance, etc.) changes accordingly, which will cause the network topology to change rapidly and seriously affect the network performance.

Table 1. Comparison of MANET, VANET and FANET.

Characteristics	MANET	VANET	FANET			
			Small UAVs		Large UAVs	
			FW *	RW *	FW *	RW *
Node Speed	Low	Medium	Medium	Medium	High	High
Density	High	High	Low			
Power Consumption	Low	High	Med	Med	High	High
Network Connectivity	High	Medium	Low			
QoS	Medium	Depends on the applications				
Mobility Patterns	2D	2D	3D			
Mobility Degree	Low	Medium	Low-Med	Low-Med	Med-High	Med-High
Altitude	Close to ground	Close to ground	Low	Low	High	High
Static Pause	Yes	Yes	No	Yes	No	Yes
Topology Variation	Occasionally	Frequently	Very frequently		Very frequently	
scalability	Medium	High	Low			
Typical Mobility Models	regular	Restricted through roads	Depends on the applications and need to pre-determined			

* FW: fixed-wing, * RW: rotary wing.

The main research object of this paper is highly dynamic FANETs. Specifically, it corresponds to the FANETs listed in Table 1 in which the nodes move fast and the network topology changes drastically.

The characteristics of the FANETs discussed in this paper are: small and medium-sized networks, high mobility nodes far from the ground at high altitudes, sufficient node energy and frequent network topology changes.

In practical applications, FANETs encounter different application scenarios when nodes complete complex tasks. Nodes move in different ways in different scenarios, which causes large changes in the network topology. Traditionally, nodes cannot perceive this change, nor can they adapt quickly to it, so the network performance will deteriorate dramatically. The main purpose of this paper is to find a method by which nodes in a network can perceive changes of topology and adjust the routing strategy appropriately when the topology changes greatly, so that the network performance could be maintained at a high level. The most concerned network performance is the packet delivery ratio, followed by the average end-to-end delay, the average jitter, and finally the network throughput.

Many routing protocols have been proposed for mobile ad hoc networks [4]. Each protocol has its own characteristics and application scenarios. For example, the Optimized Link State Routing Protocol (OLSR) [5] is suitable for high-density MANETs, and the Destination-Sequenced Distance-Vector Routing (DSDV) [6] is more suitable for small-scale MANETs. When the scenario is complex and the network topology changes rapidly, a single routing protocol cannot meet the requirements of the variable node mobility mode, nor can network performance be guaranteed. Accurate perception of the surrounding environment and appropriate adjustment strategies of nodes have become the trends of adaptive routing protocols with the increasing requirements of quality of service (QoS). In response to the above objectives, this paper proposes a solution for choosing routing protocols in highly dynamic FANETs, namely TARCS (topology change aware-based routing protocol choosing scheme). The TARCS scheme is divided into the following steps:

Firstly, a topology change sensing method is proposed, which can measure the topology change between nodes of a high dynamic FANET, and a mobility metric TCD, the topology change degree with several levels is defined to quantify the perceived results.

Secondly, a heuristic routing strategy of FANETs is proposed using the measurement results of the sensing method. The policy compares the measured topology change result with the threshold reference value, determines the current node movement mode, and appropriately adjusts the routing protocol used in the network, so that the network routing protocol is more suitable for the changed mobile scenario, and finally the purpose of improving network performance in highly dynamic topology is achieved.

Simulations show that compared with traditional practices, the proper use of TARCS in highly dynamic MANETs in complex scenarios can effectively improve some aspects of network performance.

The content discussed in this article is based on the following assumptions:

- The node motion is simplified to two-dimension movements.
- The transmission range of each node is a circular area with the node in the center and a radius of r. The transmission power of the node is constant within this area.
- The number of nodes in the network is constant during the network lifetime.

The remainder of this paper is organized as follows. In Section 2, the TARCS is introduced in detail. In Section 3, the TARCS strategy is validated in the form of simulation experiments. The experimental results are discussed in Section 4, and Section 5 contains the conclusion.

2. Related Work

The following paper summarizes related literature from two aspects: topological change perception and topological change perception processing.

2.1. Topology Change Perception Methods and Processing Methods

The state-of-the-art perceptions of topology changes are limited to mobility aware of nodes and paths. Radhika Ranjan Roy [7] described mobility models and mobility metrics in detail. Bai Fan et al. [8] defined several mobile metrics, such as node spatial dependence, node temporal dependence and geographic restrictions to capture mobility characteristics, and proposed the IMPORTANT framework for analyzing the impact of mobility on the performance of routing protocols. Jie Hong et al. [9] explained the relationship among node mobility, network topology changes and network performance such that the high speed of nodes in a MANET did not mean rapid topology changes, and changes in network topology caused by changes of node speed and direction ultimately affected network performance. A dynamic routing algorithm improved from the dynamic source routing algorithm (DSR [4]) and proposed by Ehssan S. et al. [10] by selecting relatively static nodes according to the Doppler frequency shift was applied to the aeronautical ad hoc networks (AANETs) with high-speed nodes. Zheng, Y. et al. [11] proposed the mobility and load aware OLSR (ML-OLSR) protocol, which combined mobile sensing and load sensing with the procedures of selecting more stable nodes as the multi point relay (MPR) points, and lower-load routing to reduce the average end-to-end delay. The protocol is applicable to the unmanned aerial vehicle (UAV) networks with high-speed nodes and unbalanced loads. The protocol proposed by Zhou J.H. et al. [12] was also applicable to AANETs in which the effective time of the largest link was estimated according to the node relative speed and the Doppler shift of the received signal; meanwhile the traffic load was referenced during the route discovery. The mobility pattern aware routing protocol proposed by Hung, C-C. et al. [13] requires geographic assistant location information and relies on IEEE 802.16 base station assistance for a heterogeneous vehicular network (HVN). Athanasios B. et al. [14] proposed a protocol framework with the defined three-layer services of mobility classification service (MCS), strategy selection service (SSS) and routing service (RS) to make nodes with different mobility levels use different routing protocols. It is suitable for high-speed mobile networks with sparse nodes. Yu, Y. et al. [15] combined the DSR protocol with the ant colony optimization algorithm (ACO), judging the stability between nodes and neighbors according to the received signal power, using the ACO to detect the network congestion degree. Nodes with high mobility in the cache were deleted according to the node stability and the congestion degree. The premise of the protocol proposed by Swidana, A. et al. [16] was that the velocity of the node had been measured before, the basic idea of which was to prevent high mobility nodes from participating in route discovery. Khalaf, M. et al. [17] proposed two probability models for speed perception, with which two protocols were proposed by improving the ad-hoc on-demand distance vector (AODV) [18] routing algorithm. The protocol proposed by Moussaoui, A. et al. [19] judged node stability based on the received signal power, and improved the OLSR protocol by selecting the more stable neighbor node as the MPR node. Brahmbhatt, S. et al. [20] proposed a protocol to select the most powerful node from the received signal as a stable link in light of signal strength-based link stability estimation to solve the routing failure problem in multipath networks.

2.2. Discussion

The mobility aware methods noted above can be classified into the following categories:

- Neighbor node or link mobility awareness [10–17,19,20].
- Neighbor node mobility awareness combined with congestion level perception [11,12,15].
- Neighbor node mobility trend prediction [12].

These mobility aware methods can also be divided into location aided [11,13,14,16,17] and non-location aided [10,12,15,19,20]. The former refers to estimating the neighbor node/path mobility by means of information from auxiliary devices, such as global positioning system (GPS) receivers, with the advantage of simple and fast calculation and the disadvantage of introducing GPS errors and new interference. The latter usually measures the position or speed of the neighbor node according to the power or the Doppler shift of the received signal. Although no error or other interference is introduced, the direction and distance of the neighbor node are not accurately determined.

The processing method after mobility awareness can be summarized into the following:

- Selecting stable node/link during route discovery [10–13,17,19].
- Excluding or preventing high mobility nodes from participating in route discovery [15,16].
- Node mobility classification and decision making [14].

The mobility awareness methods, the processing methods and the proposed TARCS are summarized in Figure 2.

However, the methods in the above literature are only limited to measuring node mobility or link stability caused by mobility, without involving accurate definition and accurate measurement of topology changes.

In this paper, the methods of perception and perceived result processing are different. In terms of the sensing method, the above literature uses only a single parameter (velocity or distance) for perception, but TCD is used here to quantify the topological changes between nodes from multiple aspects, including distance, velocity, direction and neighbors' number, and to accurately reflect the topological changes. As to the subsequent processing method, in this paper, after determining the node's mobility mode based on the perceived result, the routing protocol is re-selected according to the current mobility mode characteristics. Essentially this is a routing protocol optimization selection strategy.

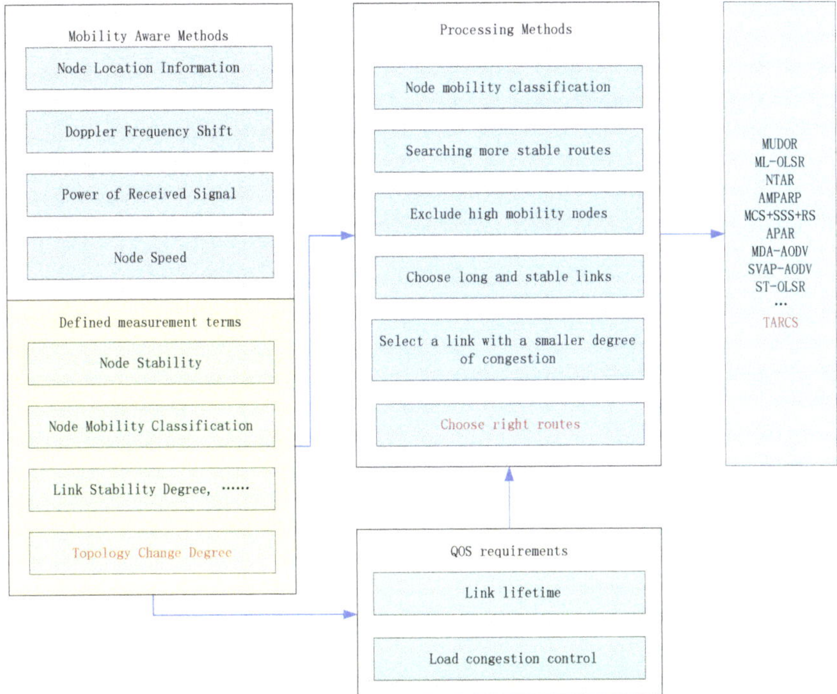

Figure 2. The mobility aware methods and the subsequent processing methods.

3. A Topology Change Aware Routing Protocol Choosing Scheme (TARCS)

A topology change aware routing protocols choosing scheme (TARCS) is proposed based on the results of the periodic topology change awareness (PTCA) and the adaptive route choosing scheme (ARCS). The purpose of the PTCA is to accurately measure topology changes and determine the movement mode. The purpose of the ARCS is to select an appropriate routing protocol for the current motion mode and to ensure that network performance is affected as little as possible. The periodic topology change perception is based on the topological change degree (TCD), a mobility indicator proposed by the previous study, which could measure the topology changes of nodes and distinguish the Random Waypoint model (RWP [7]), the Reference Point Group Mobility Model (RPGM [7,21]) with different group numbers and the Pursue model [7,21].

3.1. Description

The topology change aware routing protocol choosing scheme (TARCS) includes two procedures: the periodically topology change aware (PTCA) and the adaptive route choosing scheme (ARCS). Accurate perception of topology changes is a prerequisite for routing protocol adjustment and reselection. The routing protocol selection is based on a preset topology change threshold/interval value.

Traditional routing protocols either maintain routing paths during node movement (active routing) or start routing discovery when nodes need to communicate (passive routing). Neither active routing nor passive routing pays close attention to the surrounding topology. This is completely sufficient for low-speed nodes, but for high-speed nodes the topology changes so fast that nodes cannot respond to the rapidly changing environment and adjust strategy if they lack the knowledge of the external environment. Obviously, the consequence is that the nodes are not well adapted to the topology changes, the network performance is severely constrained, and the task cannot be successfully completed.

The advantage of TARCS is that:

- It can monitor the topology changes around the nodes in real time.
- Nodes will respond to the perceived results in time and adjust to select the appropriate routing protocol.
- It will make the current routing protocol more suitable for node mobility mode and topology change requirements.

Assume that node A (n_a) will send information to node B (n_b) in a FANET. Figure 3 shows the comparison of communication mode from n_a to n_b with and without the TARCS. Figure 3a shows that n_a communicates with n_b directly through a routing protocol in the traditional manner. n_a first establishes a valid path with n_b through route discovery, then sends data and maintains the route. Figure 3b shows the communication mode from n_a to n_b after using the TARCS.

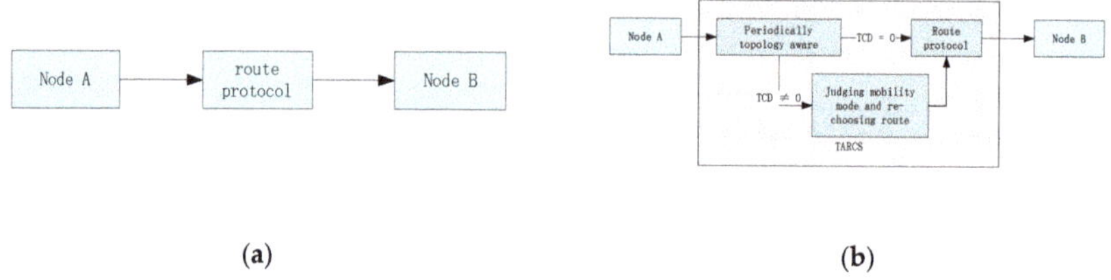

(**a**)   (**b**)

Figure 3. Communication mode comparison between two nodes without and with the topology change aware routing protocol choosing scheme (TARCS). (**a**) Traditional communication mode between nodes without the TARCS; (**b**) communication mode between nodes with the TARCS.

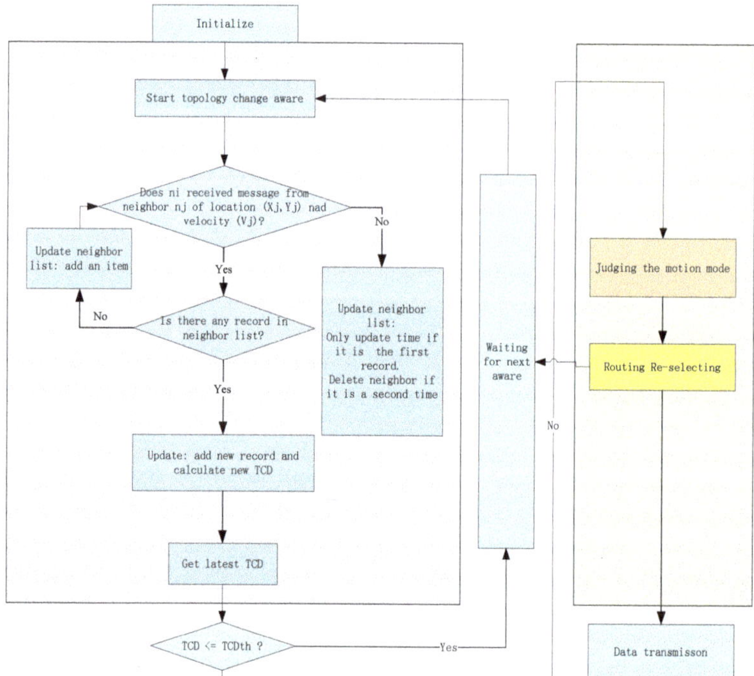

Figure 4. Process flow of the TARCS.

Figure 4 shows the basic workflow of the TARCS. The TARCS is divided into two phases. The left region represents the periodic topology change aware (PTCA), and the right region represents the adaptive routing protocol selection (ARCS).

3.2. Periodically Topology Change Aware (PTCA)

The purpose of PTCA is to periodically perceive the topology changes of each node and its neighbors within a one-hop transmission range in the network. Since the movement of nodes causes the change of the relative position between the neighboring nodes and in the number of neighboring nodes, thereby causing the network topology to change, the discussion of the topology change begins with the mobility of the node.

Figure 5 shows the topology change caused by the change in distance among nodes. Black dots and red dots represent moving nodes, and dashed circles indicate the effective transmission range of nodes. Figure 5a shows the topology in the initial state. Nodes n_1 and n_2 are within one hop of n_0. Node n_3 is within one hop of n_1, and n_4 is within one hop of n_3. Node n_0 establishes a route through n_1, n_3, and n_4. From this moment on, n_1 starts moving down. Figure 5b shows the network topology after time T. At this time, n_1 has moved out of the effective range of n_0, and the distance between n_0 and n_1 increases. If n_0 still sends data to n_4, it can only pass the path: $n_0 \rightarrow n_2 \rightarrow n_3 \rightarrow n_4$. It can be seen that topology changes could be caused by distance changes between nodes.

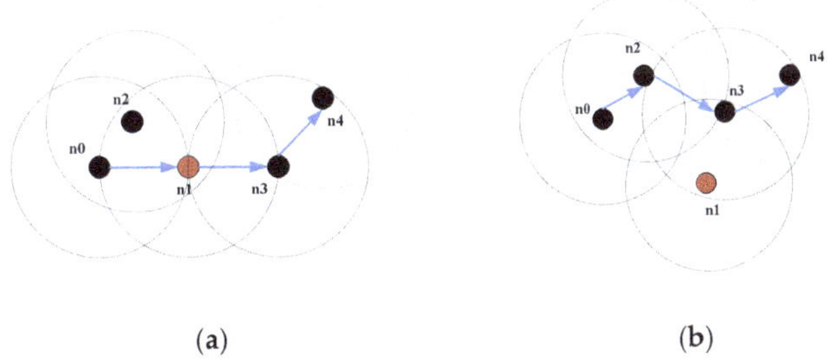

(a)   (b)

Figure 5. Topology change caused by nodes distance change: (**a**) topology in the initial state; (**b**) topology after period T.

Figure 6 shows the topology change caused by changes in the direction of movement between nodes. Figure 6a is the initial state. From this moment on, n_0 will move around n_1. After time T, the distance between n_0 and n_1 is still the same. However, there is no effective path between n_0 and n_4 due to the change of direction between n_1 and n_0. As is shown in Figure 6b. It can be seen that changes in the direction of movement of nodes can also cause topology changes.

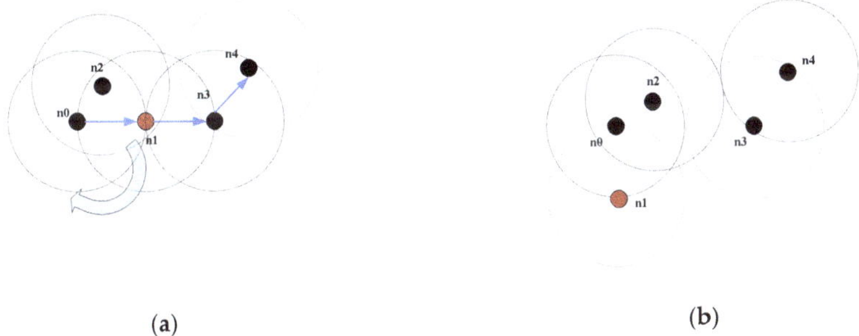

(a)   (b)

Figure 6. Topology change caused by nodes direction change: (**a**) topology in the initial state; (**b**) topology after time T.

Figure 7 shows the topology change caused by changes in the relative rate between nodes. Figure 7a is the initial state. All nodes move up, but the nodes move at different rates. The rate of n_1 is less than the rate of the others. Figure 7b is the topology after time T. It can be seen that the topology has also changed.

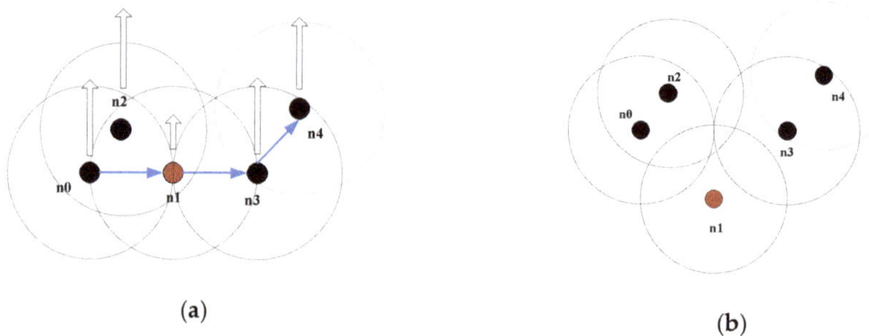

(a)   (b)

Figure 7. Topology change caused by a difference of rate: (**a**) topology in the initial state; (**b**) topology after T time.

Figure 8 illustrates the topology change caused by the number of neighbors. Figure 8a is the initial state. n_1 and n_2 are both one-hop neighbors of n_0. The initial path of n_0 to n_4 is $n_0 \rightarrow n_1 \rightarrow n_3 \rightarrow n_4$. In Figure 8b n_1 exits the network for some reason (e.g., hardware failure or energy exhaustion) after time T. At this time, the number of neighbors of n_0 is reduced resulting in no complete path between n_0 and n_4. Therefore, changes in the number of neighbor nodes can also cause topology changes, but this condition is not sufficient. For example, although n_2 is also a neighbor of n_0, the departure of n_2 does not affect the path from n_0 to n_4.

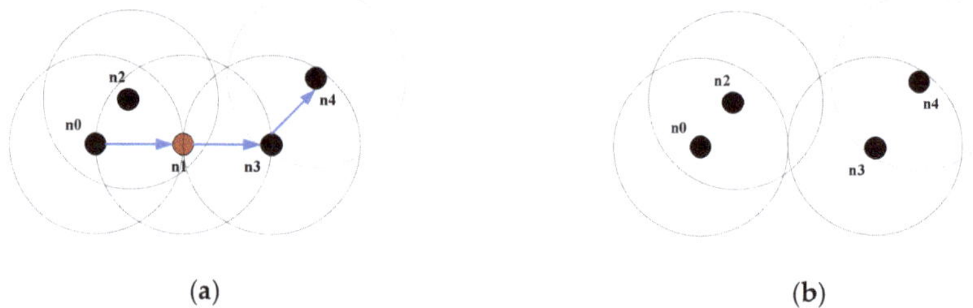

(a)   (b)

Figure 8. Topology change caused by a change in the number of neighbors: (**a**) topology in the initial state; (**b**) topology after time T.

From the analysis above, it is concluded that each of these factors, namely, the relative position, motion direction, relative rate, and number of neighbor nodes, may cause a change in the local topology between neighboring nodes. Thus, a new mobility metric, the topology change degree (TCD) was defined by selecting parameters such as node distance, motion direction and movement rate, to measure the topology change between neighbor nodes. The main purpose of TCD is to accurately quantify topology changes between neighboring nodes and quickly identify different motion modes.

Topology change degree between nodes i and j within time T, the $TCD_{i,j}$, represents the degree of topology change between the node i and its neighbor node j in the T time period. The detailed definition is shown in Equation (1).

$$TCD_{i,j}(t, t+T) = \left(w_1 \cdot \frac{\left| d_{i,j}^{t+T} - d_{i,j}^{t} \right|}{d_{i,j}^{t}} + w_2 \cdot \frac{\left| \theta_{i,j}^{t+T} - \theta_{i,j}^{t} \right|}{2\pi} + w_3 \cdot \frac{\left| relv_{i,j}^{t+T} - relv_{i,j}^{t} \right|}{relv_{i,j}^{t}} \right) \tag{1}$$

where:

- $\dfrac{\left|d_{i,j}^{t+T} - d_{i,j}^{t}\right|}{d_{i,j}^{t}}$ indicates the change in distance between node i and node j;

- $\dfrac{\left|\theta_{i,j}^{t+T} - \theta_{i,j}^{t}\right|}{2\pi}$ indicates the change of direction between node i and node j;

- $\dfrac{\left|relv_{i,j}^{t+T} - relv_{i,j}^{t}\right|}{relv_{i,j}^{t}}$ indicates the change of relative rate between node i and node j;

- ω is the weight coefficient: here $\omega_1 = \omega_2 = \omega_3 = 1$.

Average topology change of node i and its neighbors in time T, is define as the $TCD_{i,nbrs}(t, t+T)$. Suppose the number of neighbors j of node i is n at time T ($1 \leq j \leq n$), then the average topology change degree of node i and neighbors in time T is defined as the topology change degree average value of each pair of node i and neighbor j, as defined in Equation (2).

$$TCD_{i,nbrs}(t, t+T) = \frac{\sum_{j=1, j\neq i}^{n} TCD_{i,j}(t, t+T)}{n} \tag{2}$$

The whole network topology change in time T, the $TCD_{ntwrk}(t, t+T)$, is defined as the sum of the topological changes of all nodes in the network if the total number of nodes in the network is N ($N > n$), as shown in Equation (3).

$$TCD_{ntwk}(t, t+T) = \frac{\sum_{i=1}^{N} TCD_{i,nbrs}(t, t+T)}{2} \tag{3}$$

Experiments have shown that the difference in node mobility affects the network TCD. In addition, TCD has been confirmed to reflect the topology changes around the nodes and the network, and can distinguish several different mobility modes, including RWP and RPGM (with different groups).

The first step of TARCS is to perform periodic topology change perception on each node of the entire network to obtain accurate topology change perception results. The main method of node mobility mode discrimination of this paper is to compare the topology change perception result TCD_{ntwk} and the topological change threshold reference value TCD_{TH}.

3.3. Adaptive Routing Choosing Scheme (ARCS)

The next step of TARCS is to perform adaptive routing selection, ARCS, after obtaining the results of the surrounding topology changes. The ARCS includes node mobility mode discrimination and routing protocol selection. The judgement role is:

$$\begin{cases} Routing\ protocol\ A, if\ TCD_{ntwrk} \leq TCD_{TH} \\ Routing\ protocol\ B, if\ TCD_{ntwrk} > TCD_{TH} \end{cases} \tag{4}$$

In practical applications, the TCD_{TH} can be pre-calculated with reference to a specific moving model. Experiments in Section 4 show the specific calculation process of the threshold value. It is also possible to set a threshold range for discrimination. If the perceived result TCD_{ntwk} is within the allowed interval, the current route protocol is continued to be used, and if the perceived result exceeds the threshold interval, the routing protocol selection is restarted.

Routing is based on the appropriate mobility model for each routing protocol. Therefore, it is necessary to calculate and classify possible scenarios, each mobility model with its TCD_{TH}, and the corresponding routing protocols in advance. After the routing protocol is selected, nodes communicate through the newly set routing protocol until the next topology change perception.

4. Evaluations and Results

Simulation experiments are designed to evaluate the scheme. It is assumed that all nodes can receive signals from neighboring nodes within one hop in each direction, the node energy is sufficient, and the transmission distance is constant. The speed of the nodes, the distance between nodes, and the direction of motion of the nodes are known. Nodes move only in two dimensions.

In order to more clearly reflect the highly dynamic network topology, the Chain scenario [22] is selected to simulate the highly dynamic scene, which combines RWP, RPGM and Pursue model to simulate nodes to complete a series of tasks such as search/exploration, reconnaissance/patrol and target tracking/rescue. The duration is 600 s. From 0 s to 200 s nodes move with the RWP model, then they move with the RPGM (g = 5) model from 200 s to 400 s, and finally nodes move with the Pursue model.

The simulation tool is NS-3.25 [23] (Network Simulator 3, Version 3.25). The region is a square of 2000 m × 2000 m. The number of nodes is 50 with 50 m/s as the lower speed and the 500 m/s as the higher speed. The Free space propagation model [23,24] is selected as the propagation mode and the constant speed propagation delay model is selected as the delay model. The transmission mode is constant bit rate (CBR) with a bit rate of 16 Kbps and a packet size of 1024 bytes. The MAC (media access control) protocol is IEEE 802.11g. The alternative routing protocols involved are AODV, OLSR, and DSDV. A detailed description of each mobility model and routing protocol can be found in the references. Table 2 lists the experimental parameter items and parameter values.

Table 2. Parameters of simulation experiments.

Parameters	Values
Simulation region	2000 m × 2000 m
Node number	50
Mobility Model	Chain: RWP + RPGM (g = 5) + Pursue
Simulation Time	602 s
Node speed	50 m/s, 500 m/s
Route protocol	OLSR / AODV / DSDV
Propagation model	Free space propagation model
Delay model	constant speed propagation delay model
Data mode	CBR
Packet size	1024 bytes
Data rate	16 Kbps
MAC protocol	IEEE 802.11bDCF
Transmission range	140 m

In the first experiment, the network performance is compared with or without the TARCS scheme. The second experiment is to compare the impact on the network performance with different TARCS strategies.

4.1. Validation of TARCS

This experiment mainly verified the difference in network performance with or without TARCS. The aware interval is set to 50 s, which means that the topology is perceived 13 times with each rate. Since the TCD_{TH} are different with different node numbers and node rates, it is necessary to preliminarily set the TCD_{TH} according to the node rate and the specific mobility model.

In this experiment, when node speed is 50 m/s the TCD_{TH} is set as follows:

$$\begin{cases} TCD_{TH}(RWP, RPGM(g=5)) = 150 \\ TCD_{TH}(RPGM(g=5), Pursue) = 1000 \end{cases} \tag{5}$$

The meaning of $TCD_{TH}(RWP, RPGM(g = 5))$ is the threshold of the network TCD distinguishing from motion with the RWP mode and the RPGM (g = 5) mode. Assuming TCD_{ntwrk} is the newly acquired TCD value of the whole network, the criteria are:

$$mobility\ model = \begin{cases} RWP & if\ TCD_{ntwrk} < TCD_{TH}(RWP, RPGM(g = 5)) \\ RPGM(g = 5) & if\ TCD_{ntwrk} > TCD_{TH}(RWP, PGM(g = 5)) \end{cases} \tag{6}$$

Similarly, $TCD_{TH}(RPGM(g = 5), Pursue))$ represents the TCD threshold value that can distinguish between motion in RPGM (g = 5) mode and in Pursue mode.

When node speed is 500m/s the TCD_{TH} is configured as follows:

$$\begin{cases} TCD_{TH}(RWP, RPGM(g = 5)) = 1000 \\ TCD_{TH}(RPGM(g = 5),\ Pursue) = 2000 \end{cases} \tag{7}$$

And the criteria are:

$$mobility\ model = \begin{cases} RPGM(g = 5) & if\ TCD_{ntwrk} < TCD_{TH}(RPGM(g = 5), Pursue) \\ Pursue & if\ TCD_{ntwrk} > TCD_{TH}(RPGM(g = 5), Pursue) \end{cases} \tag{8}$$

The calculation method for the threshold value will be introduced in Section 4.3.

According to the conclusion of [5], this experiment uses the following routing protocol selection scheme: the AOPV protocol is used in the RWP scenario, the AODV protocol is used in the RPGM scenario, and the DSDV protocol is used in the Pursue scenario. The TCD perception result at time t is represented by $TCD(t)$. The perceptual moment, the actual movement model, the topological change perception result $TCD(t)$, the judged mobility model, and the selected routing protocol are listed in Table 3.

Table 3. The topological change degree (TCD) perception results and route protocol choosing scheme of Experiment 1.

Perceptual Moment	Actual Mobility Model	Node Speed = 50 m/s		Node Speed = 500 m/s		Route Protocol
		TCD(t)	Judged model	*TCD(t)*	Judged model	
0	RWP	0	RWP	0	RWP	AODV
50	RWP	2.078	RWP	0	RWP	AODV
100	RWP	15.459	RWP	842.592	RWP	AODV
150	RWP	34.364	RWP	782.536	RWP	AODV
200	RPGM (g = 5)	177.845	RPGM	589.106	RWP	AODV
250	RPGM (g = 5)	560.596	RPGM	2513.24	Pursue	DSDV
300	RPGM (g = 5)	658.254	RPGM	678.321	RWP	AODV
350	RPGM (g = 5)	493.418	RPGM	2102.88	Pursue	DSDV
400	Pursue	1963.96	Pursue	4144.1	Pursue	DSDV
450	Pursue	974.397	RPGM	2042.52	Pursue	DSDV
500	Pursue	3800.78	Pursue	2312.88	Pursue	DSDV
550	Pursue	2332.39	Pursue	4482.67	Pursue	DSDV
600	Pursue	4786.97	Pursue	3666.45	Pursue	DSDV

The perceptual results show that among the 26 perceptual moments, the perceived result $TCD(t)$ is less than the threshold value and the determined mobility model does not match the actual model only for the speed of 50 m/s at 450 s and the speed of 500 m/s at 250 s, 300 s and 350 s. The rest of the perceived results and judgment results are correct. The rest of the perception results are correct, within the threshold range, and the judgment results are consistent with the actual movement model. The perceived efficiency is 84.6%. The above results show that TCD is effective for network topology change perception, and that the granularity setting and the TCD_{TH} setting are very important for the judgment.

The cause of the error is likely to be the following factors. The threshold setting is not reasonable enough, and the time of each mobility model is not long enough, so the mobility model is not very stable when the nodes move at high speed.

Figure 9 illustrates the comparisons of performance metrics, including the packet delivery ratio, the network throughput, the average end-to-end delay and the average jitter under two different node speeds (50 m/s and 500 m/s) with the TARCS (assume that all mobility models are accurately identified)and with other routing protocols.

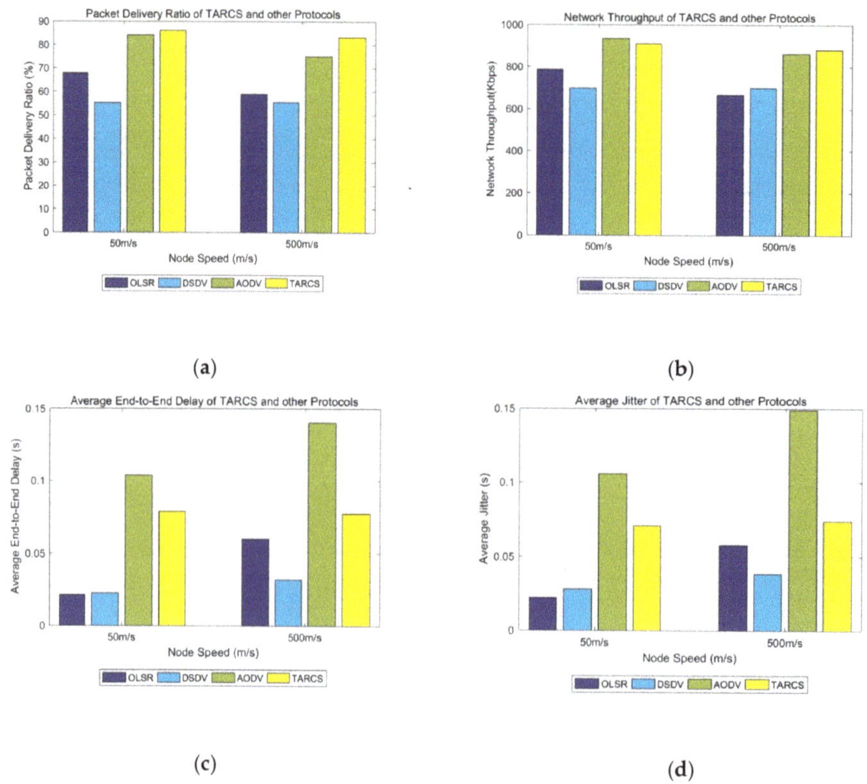

<table>
<tr><td>(a)</td><td>(b)</td></tr>
<tr><td>(c)</td><td>(d)</td></tr>
</table>

Figure 9. Performance comparisons with and without the TARCS: (**a**) packet delivery ratio; (**b**) network throughput; (**c**) average end-to-end delay; (**d**) average jitter.

It can be seen from Figure 9a that when node speed is 500 m/s, the packet delivery ratio is 58.9435% only with OLSR, 55.4443% only with DSDV, 74.9448% only with AODV, and rises to 83.3388% with the TARCS.

Figure 9b shows the network throughput comparison with and without TARCS under different node speeds. Taking the speed of 500 m/s as an example, the network throughput is 666.716 Kbps when only OLSR is used, 701.095 Kbps when DSDV is only used, 862.448 Kbps when AODV is only used, and 881.438 Kbps when the TARCS scheme is used. Network throughput is increased by 32.2%, 25.7%, and 2.2%, respectively, compared to using only OLSR, DSDV, and AODV.

Figure 9c shows the average end-to-end delay comparison. Regardless of the rate, the average end-to-end delay is different from the first two performance metrics. Using the TARCS strategy chosen for this experiment, the average end-to-end delay of the network is not the smallest, but between the highest 0.141 s (AODV) and the lowest 0.032 s (DSDV), and is also higher than the OLSR of 0.06 s with the node speed of 500 m/s.

The comparison result of the average jitter is similar to that of the delay, as is shown in Figure 9d. It is between the highest value of 0.149 s (only with AODV) and the lowest value of 0.0383 s (only with DSDV) with the node speed of 500 m/s.

The reason why the advantage of TARCS is not obvious when the speed is low is that the topology change of the network is not obvious when the node speed is low.

The results of first experiment show that:

- In a complex scenario, using the TARCS strategy will have a certain impact on network performance.
- Using the same TARCS strategy at different node speeds has a different impact on network performance.
- In the TARCS strategy, the setting of TCD_{TH} is very important and should be performed in advance.

4.2. Impact Analysis of Different TARCS Strategies on Network Performance

The second experiment is to illustrate different routing strategies can affect different aspects of network performance. This experiment uses three different routing strategies (see Table 4 for details), and the rest of the parameters are the same as Experiment 1. The first strategy is AODV→AODV→DSDV, i.e., in the RWP model the AODV protocol is used, in the RPGM model the DSDV is used, and in the Pursue model the DSDV protocol is used. It is represented by TARCS1 in Table 3. Similarly, TARCS2 is OLSR→AODV→DSDV and TARCS3 is AODV→DSDV→DSDV.

Table 4. Three different schemes of TARCS.

Mobility Model	TARCS1	TARCS2	TARCS3
RWP	AODV	OLSR	AODV
RPGM (g = 5)	AODV	AODV	DSDV
Pursue	DSDV	DSDV	DSDV

The network performance under three different types of TARCS strategies is shown in Figure 10. The histogram on the left side of each figure shows the performance for the node speed of 50 m/s, and the right side shows the performance for the speed of 500 m/s.

It can be seen from Figure 10 that when the node speed is 50 m/s, the packet delivery ratio of the three strategies has little difference and TARCS1 is slightly dominant (TARCS1: 86.1261%, TARCS2: 77.3794%, and TARCS3: 69.3387%). In terms of network throughput, TARCS1 (910.061 Kbps) and TARCS2 (909.755 Kbps) are slightly superior to TARCS3 (766.448 Kbps). In terms of average end-to-end delay, TARCS3 (0.0495 s) is significantly better than TARCS1 (0.0789 s) and TARCS2 (0.05578 s). Regarding average jitter, TARCS3 (0.0483 s) is better than the other two (TARCS1: 0.0708 s and TARCS2: 0.05538 s respectively).

When the node speed is increased to 500 m/s, the packet delivery ratio of TARCS1 is the highest (sorted in descending order: TARCS1 83.3388%; TARCS3 73.0154%; and TARCS2 69.5115%). The network throughput of TARCS1 is also the highest of the three (sorted in descending order: TARCS1, 881.438 Kbps; TARCS2, 821.2191 Kbps; and TARCS3, 813.157 Kbps). Yet, in terms of average end-to-end delay, TARCS3 has the lowest value (sorted in ascending order: TARCS3, 0.04968 s, TARCS1, 0.07751 s; and TARCS2, 0.1221 s). The average jitter of TARCS3 is also the lowest (sorted in ascending order: TARCS3, 0.04838s; TARCS1, 0.07397 s; and TARCS2, 0.1192 s). None of the network performances is effectively improved by TARCS2.

The results of the second experiment indicate that using different TARCS strategies can affect different network performance metrics. The indicators affected is related to the characteristics and the appropriate scenarios of each protocol [5]. The AODV protocol is more suitable for RWP scenarios, while DSDV is more suitable for RPGM (g = 1) scenarios. OLSR sits between these two and is very sensitive to node mobility speed and has higher jitter. The experimental results above also show this feature. Overall, in order to determine which strategy will get the best performance, in addition to the network performance requirements, the network designer must also deeply understand the characteristics of the selected protocol and the suitable scenarios, and make an accurate judgment on the moving scene of nodes.

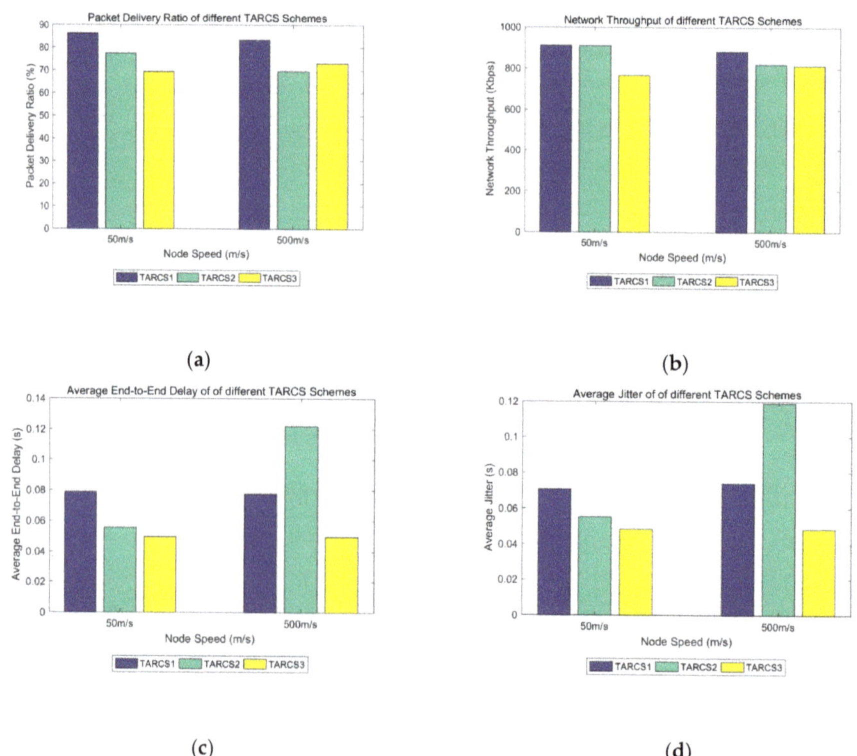

(a)

(b)

(c)

(d)

Figure 10. Performance comparisons of different TARCS: (**a**) packet delivery ratio; (**b**) network throughput; (**c**) average end-to-end delay; (**d**) average jitter.

4.3. Validation of The Topology Change Awareness Method

The experiment in this section is intended to discuss the validity of the topology change awareness method and to explain the calculation of the topological change threshold TCD_{TH}.

We have selected several different forms of node motion, which are RWP, RPGM (g = 50), RPGM (g = 25), RPGM (g = 10), RPGM (g = 5) and Pursue. g represents the group number.

The simulation region is still 2000 m × 2000 m. The duration is 950 s. the number of nodes is 50. The TCD_{ntwrk} of different motion patterns is shown in Figure 11.

From Figure 11a, it can be seen that:

- The TCD_{ntwrk} value is different when nodes move in the different patterns. As the number of groups decreases, the value of TCD_{ntwrk} gradually increases.
- In the same motion mode, the change of node speed has little effect on the TCD_{ntwrk} value.

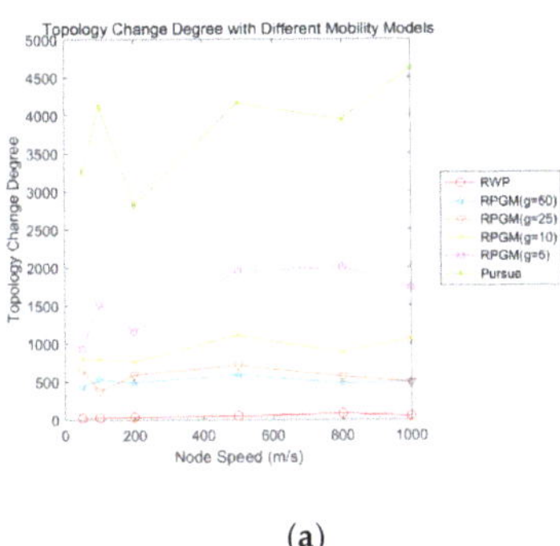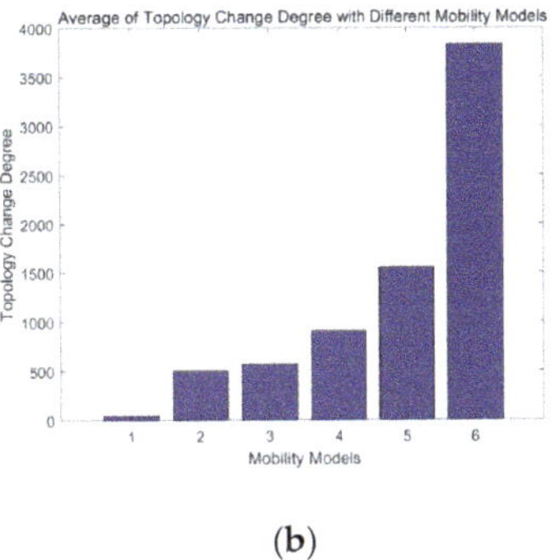

(a)   (b)

Figure 11. Comparisons of topology change degree values with different mobility patterns: (a) TCD_{ntwrk}; (b) the average of TCD_{ntwrk} (1-RWP, 2-RPGM (g = 50), 3-RPGM (g = 25), 4-RPGM (g = 10), 5-RPGM (g = 5), 6-Pursue).

The explanation for why the TCD_{ntwrk} value of RWP model is the smallest and that of Pursue model is the largest depends on how the nodes move in these models.

Under the premise of the same area, the same number of nodes and the similar speed, the TCD_{ntwrk} of the RWP model is smaller than that of the RPGM model and of the Pursue model, which is caused by the motion pattern of the nodes in the RWP model. When moving in the RWP mode, the nodes are randomly distributed in the active area. Generally, the area of the active area is much larger than the transmission range of the nodes. In this case, the motion of the node belongs to the individual motion, so the number of neighbors is small and not fixed, thus the value of each $TCD_{i,j}$ is small, and the values of $TCD_{i,nbrs}$ are also small.

When nodes move in groups, such as RPGM (g = 5), every 10 nodes are gathered into one group, and each group moves in an overall manner. The number of neighbors of a node is fixed and may increase instantaneously. Because the relative motion of the nodes inner group is weak, and the inter-group relative motion is frequent, when different groups meet and are interlaced, the number of neighbor nodes may increase instantaneously.

When nodes move in the manner of Pursue, all nodes are gathered into one group. The number of neighbors of each node is large and fixed, and the node is usually connected with its neighbors, so the $TCD_{i,nbrs}$ value are larger.

Figure 11b shows the average value of each mobility model in Figure 11a. The average value of TCD_{ntwrk} of each mode is more likely to show this difference. A feasible threshold is calculated by averaging the TCD_{ntwrk} average values of the two modes of movement that need to be compared. The threshold values in the above experiments are calculated by this method.

4.4. Analysis of the Influence of Node Density on TCD

In this section, the impact of changes in the number of nodes on TCD results will be analysed.

We believe that the number of nodes should be considered together with the range of node activity, which is reflected in the indicator of node density. Here, the node density in the network is defined as the number of nodes per unit area within the effective area of the node's movement in the network.

$$Node\ Density = \frac{Node\ Number}{S_{region}} \tag{9}$$

The revised version gives comparisons of the TCD_{ntwrk} when the node density changes. In the case of group movement, the nodes are relatively concentrated and follow the reference point motion. Taking the RWP model as an example, we analyze the change of TCD_{ntwrk} when the number of nodes changes. There are several cases.

Case 1: The area of simulation region is fixed, and the number of nodes changes (the node density changes). We take the RWP model as an example. Figure 12a shows the result. It can be seen that in the same area of region, the value TCD_{ntwrk} increases as the density of nodes increases. This phenomenon can be explained by Equation (2). As the node density increases, the number of node neighbors also increases.

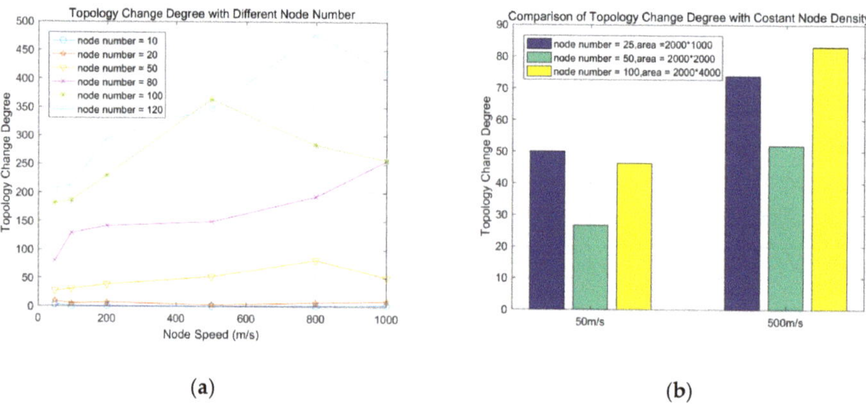

Figure 12. Comparisons of TCD_{ntwrk} with changed node density and fixed node density: (**a**) the node density is variable; (**b**) the node density is fixed.

Case 2: The region is proportional to the number of nodes (the node density is constant). Here is also an example of RWP model. In Figure 12b, the node density is fixed to 0.000125. It shows that when the node density is constant, the change in TCD is not so large.

The value of TCD reflects the degree of topology change between nodes in the network. In fact, it also reflects the degree of aggregation of nodes and the degree of unified action between nodes.

It can be seen from Equations (1) to (3) that:

- For $TCD_{i,j}$, the factors that affect its value include changes in velocity, direction, and distance between nodes.
- For $TCD_{i,nbrs}$, in addition to each single $TCD_{i,j}$ value, the number of neighbor nodes within one hop is also one of the influencing factors.
- For TCD_{ntwrk}, the number of nodes in the network directly affects the value of TCD_{ntwrk}.

4.5. Disccusions

The TARCS routing scheme proposed in this paper is more suitable for complex scenarios with highly dynamic changes in network topology. The advantages are:

1. Node cognition of its surroundings. Moving nodes can sense the topology environment between itself and neighboring nodes accurately and in a timely manner, thereby improving the node's cognitive ability in regards to its environment.
2. Adaptability. Nodes can adjust the routing protocol in real time according to the perceived result, and the adaptability of network is stronger.
3. Scalability. The TARCS adaptive routing mechanism enables a FANET to adapt to changing scenarios. By modifying the topology change degree reference threshold/interval and adding alternate routing protocols, the TARCS can be extended to more scenarios and is compatible with more routing protocols.

In addition, the TARCS routing scheme proposed in this paper also proposes a new processing idea to deal with a highly dynamic MANET, namely, the topology-change-driven protocol selection framework. Compared with the proactive, reactive and hybrid routing protocols, TARCS is more adaptable. The specific description is shown in Table 5.

Table 5. T Comparisons of topology-change-driven route and other route protocols.

Types of Route Protocols	Characteristics	Advantages	Disadvantages
Proactive routing protocol	Periodically maintain routing links; suitable for small networks.	Short delay; high quality of service; long-pass link.	High energy consumption; heavy network load and high node congestion level.
Reactive routing protocol	Establish routing discovery when needed.	Low energy consumption; light routing load.	The pass-link is not guaranteed. It takes time to establish a link. Once the link is interrupted, it needs to be re-established.
Hybrid routing protocol	Combined with proactive and reactive routing. Suitable for hierarchical networks.	Using proactive routing within the cluster, reactive between clusters	Only suitable for hierarchical networks.
Topology-change-driven protocol	Reselect the appropriate route based on topology changes.	Nodes obtain information of the surrounding topology changes in time. Multiple routing protocols are used in combination to suit complex scenarios.	Dependent on reference threshold settings and specific routing strategies.

5. Conclusions

On the basis of analyzing the factors affecting the topology changes between nodes in FANETs, a mobility metric named topology change degree (TCD) is first proposed to describe the topology changes of highly dynamic FANETs. Meanwhile, a topology change awareness method is applied to measure topology changes by periodically sensing the topology changes of FANETs and to distinguish the mobility modes of nodes. Then a heuristic routing protocol scheme, TARCS, is proposed in this paper. The TARCS strategy is based on the periodic topology change sensing method, and adaptively selects appropriate routing protocols according to the comparison results of the measured topology change degree and the topology change degree threshold.

Different from the other single routing protocols proposed in other papers, the TARCS proposed in this paper is an adaptive routing strategy. It is based on the topology change perception between nodes and adaptively selects the routing protocol based on the perceived result. Essentially, it is a flexible routing protocol usage framework. Theoretical analysis and experiments show that:

- The topology change perception method can sense the topology changes of the network and distinguish several common mobility models of FANET.
- The combined use of routing protocols is more flexible and adaptable than using a single routing protocol in complex scenarios.
- Proper use of the TARCS scheme in complex scenarios can improve network performance.

Like mobile awareness, topology change perception opens the door to coping with and addressing the problems of highly dynamic FANETs topology changes. As an adaptive routing protocol framework, TARCS is only one of the applications of topology change perception and it proposes a routing implementation idea in a complex scenario of a FANET network. However, all of the above are just a preliminary framework, and there are a series of issues that need to be discussed and resolved. For example, how to effectively set the topology change perception interval? How to accurately calculate the topology change threshold? How to configure the weight factors among the influencing factors? How to deal with the problems of changes in the number of nodes? Subsequent work will continue to study in depth on the above issues.

Author Contributions: D.Z. is the project administrator. J.H. made the investigation. J.H. and D.Z. proposed the methodology. J.H. did the validation work and finished the original draft writing. D.Z. finished the review & editing writing.

Funding: This research received no external funding.

Conflicts of Interest: The authors declare no conflict of interest.

References

1. Guillen-Perez, A.; Cano, M.D. Flying Ad Hoc Network: A New Domain for Network Communications. *Sensors* **2018**, *18*, 3571. [CrossRef] [PubMed]
2. Bekmezci, İ.; Sahingoz, O.K.; Temel, Ş. Flying Ad-Hoc Networks (FANETs): A survey. *Ad Hoc Netw.* **2013**, *11*, 1254–1270. [CrossRef]
3. Oubbati, O.M.; Lakas, A.; Zhou, F.; Güneş, M.; Yagoubi, M.B. A survey on position-based routing protocols for Flying Ad hoc Networks (FANETs). *Veh. Commun.* **2017**, *10*, 29–56. [CrossRef]
4. Sarkar, S.K.; Basavaraju, T.G.; Puttamadappa, C. Routing Protocols. In *Ad Hoc Mobile Wireless Networks-Principles, Protocols, and Applications*, 2nd ed.; CRC Press: Boca Raton, FL, USA, 2012; pp. 81–126.
5. Clausen, T.; Jacquet, P. Optimized Link State Routing Protocol (OLSR), RFC 3626 (Experimental). Available online: https://www.ietf.org/rfc/rfc3626.txt.pdf (accessed on 9 October 2018).
6. Perkins, C.E.; Bhagwat, P. Highly Dynamic Destination-Sequenced Distance-Vector Routing (DSDV) for Mobile Computers. In Proceedings of the conference on Communications architectures, protocols and applications (SIGCOMM94), London, UK, 31 August–2 September 1994; pp. 234–244.
7. Radhika Ranjan, R. Random Waypoint Mobility, Reference Point Group Mobility. In *Handbook of Mobile Ad Hoc Networks for Mobility Models*; Springer: Boston, MA, USA, 2011; pp. 637–670.
8. Fan, B.; Sadagopan, N.; Helmy, A. The Important Framework for Analyzing the Impact of Mobility on Performance of Routing Protocols for Adhoc Networks. *Ad Hoc Netw.* **2003**, *1*, 383–403. [CrossRef]
9. Hong, J.; Zhang, D. Impact Analysis of Node Motion on the performance of FANET routing protocols. In Proceedings of the 14th International Conference on Wireless Communications, Networking and Mobile Computing (WiCom2018), Chongqing, China, 18–20 September 2018.
10. Sakhaee, E.; Jamalippour, A.; Kato, N. Aeronautical Ad Hoc Networks. In Proceedings of the IEEE Wireless Communications and Networking Conference (WCNC 2006), Las Vegas, NV, USA, 3–6 April 2006; pp. 246–251. [CrossRef]
11. Zheng, Y.; Wang, Y.; Li, Z.; Dong, L.; Jiang, Y.; Zhang, H. A Mobility and Load aware OLSR routing protocol for UAV mobile ad-hoc networks. In Proceedings of the 2014 International Conference on Information and Communications Technologies (ICT2014), Nanjing, China, 15–17 May 2014.
12. Zhou, J.H.; Lei, L.; Liu, W.K.; Tian, J. A simulation analysis of nodes mobility and traffic load aware routing strategy in aeronautical ad hoc networks. In Proceedings of the 2012 9th International Bhurban Conference on Applied Sciences & Technology (IBCAST), Islamabad, Pakistan, 9–12 January 2012; pp. 423–426.
13. Hung, C.-C.; Chan, H.; Wu, E. Mobility Pattern Aware Routing for Heterogeneous Vehicular Networks. In Proceedings of the 2008 IEEE Wireless Communications and Networking Conference (WCNC2008), Las Vegas, NV, USA, 31 March–3 April 2008.
14. Bamis, A.; Boukerche, A.; Chatzigiannakis, I.; Nikoletseas, S. A mobility aware protocol synthesis for efficient routing in ad hoc mobile networks. *Comput. Netw.* **2008**, *52*, 130–154. [CrossRef]
15. Yu, Y.; Ru, L.; Chi, W.; Liu, Y.; Yu, Q.; Fang, K. Ant colony optimization based polymorphism-aware routing algorithm for ad hoc UAV network. *Multimed. Tools Appl.* **2016**, *75*, 14451–14476. [CrossRef]
16. Swidana, A.; Abdelghanya, H.; Saifana, R.; Zilic, Z. Mobility and Direction Aware Ad-hoc on Demand Distance Vector Routing Protocol. *Procedia Comput. Sci.* **2016**, *94*, 49–56. [CrossRef]
17. Khalaf, M.; Al-Dubai, Y.; Min, G. New efficient velocity-aware probabilistic route discovery schemes for high mobility Ad hoc networks. *J. Comput. Syst. Sci.* **2015**, *81*, 97–109. [CrossRef]
18. Perkins, C.E.; Royer, E.M. Ad-hoc on-demand distance vector routing. In Proceedings of the Second IEEE Workshop on Mobile Computing Systems and Applications, New Orleans, LA, USA, 25–26 February 1999.
19. Moussaoui, A.; Semchedine, F.; Boukerram, A. A link QoS routing protocol based on link stability for mobile ad hoc network. *J. Netw. Comput. Appl.* **2014**, *39*, 117–125. [CrossRef]
20. Brahmbhatt, S.; Kulshrestha, A.; Singal, G. SSLSM: Signal Strength Based Link Stability Estimation in MANETs. In Proceedings of the 2015 International Conference on Computational Intelligence and Communication Networks, Jabalpur, India, 12–14 December 2015.

21. Hong, X.; Gerla, M.; Pei, G.; Chiang, C.-C. A Group Mobility Model for Ad Hoc Wireless Networks. In Proceedings of the 2nd ACM international workshop on modeling, analysis and simulation of wireless and mobile systems, Seattle, WA, USA, 20 August 1999; pp. 53–60.
22. Bonnmotion-A Mobility Scenario Generation and Analysis Tool. Available online: http://sys.cs.uos.de/bonnmotion/ (accessed on 10 October 2018).
23. NS-3 Documentation. Available online: https://www.nsnam.org (accessed on 10 October 2018).
24. Friis, H.T. A Note on a Simple Transmission Formula. *Proc. IRE* **1946**, *34*, 254–256. [CrossRef]

 *electronics*

Article

A Novel Left-Turn Signal Control Method for Improving Intersection Capacity in a Connected Vehicle Environment

Chuanxiang Ren [1,*], **Jinbo Wang** [1], **Lingqiao Qin** [2], **Shen Li** [2] and **Yang Cheng** [2]

[1] College of Transportation, Shandong University of Science and Technology, Qingdao 266590, China; wangjb@sdust.edu.cn

[2] Department of Civil & Environmental Engineering, University of Wisconsin-Madison, 1221 Engineering Hall,1415 Engineering Drive, Madison, WI 53706, USA; lingqiao.qin@wisc.edu (L.Q.); sli299@wisc.edu (S.L.); cheng8@wisc.edu (Y.C.)

* Correspondence: renchx@sdust.edu.cn; Tel.: +86-1502-005-3186

Received: 8 August 2019; Accepted: 16 September 2019; Published: 19 September 2019

Abstract: Setting up an exclusive left-turn lane and corresponding signal phase for intersection traffic safety and efficiency will decrease the capacity of the intersection when there are less or no left-turn movements. This is especially true during rush hours because of the ineffective use of left-turn lane space and signal phase duration. With the advantages of vehicle-to-infrastructure (V2I) communication, a novel intersection signal control model is proposed which sets up variable lane direction arrow marking and turns the left-turn lane into a controllable shared lane for left-turn and through movements. The new intersection signal control model and its control strategy are presented and simulated using field data. After comparison with two other intersection control models and control strategies, the new model is validated to improve the intersection capacity in rush hours. Besides, variable lane lines and the corresponding control method are designed and combined with the left-turn waiting area to overcome the shortcomings of the proposed intersection signal control model and control strategy.

Keywords: traffic signal control; shared lane; control strategy; vehicle-to-infrastructure; variable lane line

1. Introduction

Due to the rapid increase in population and the number of vehicles, traffic congestion has had a major impact on day-to-day life. Many scholars have done a great deal of research on improving traffic congestion, such as on bus priority strategies [1,2], mass rapid transit [3,4], parking management [5,6], intelligent transportation systems [7,8], etc. As an important component of intelligent transportation systems, the efficient operation of intersections has several advantages for the transportation in a city. At an at-grade intersection, the left-turn lane and its corresponding signal phase should be set when the left-turning vehicles are sufficient [9]. The purpose of this operation is to reduce traffic conflicts and improve safety. Figure 1a shows a traditional four-leg at-grade intersection model with an exclusive left-turn lane controlled by a fixed-time system, and the signal control phases are shown in Figure 1b. In this intersection model, there are three lanes on the main street: one exclusive left-turn lane, one through lane, one shared through-right lane. There are two lanes on the minor street: one exclusive left-turn lane and one shared through-right lane. The exclusive left-turn lane is only used for left-turning vehicles. In order to realize the establishment of the left-turn lane, there are clear lane direction arrow markings in each lane, marked on the surface of the lane. Once the lane direction arrow marking is set, it cannot be changed at any time. This paper calls it static lane direction arrow marking. It causes the spatial separation of left-turning vehicles from through vehicles. Simultaneously, this arrangement is usually equipped with a left-turn signal phase duration to temporally separate left-turning vehicles from through vehicles. This intersection model with static lane direction arrow markings is applied widely in many cities around the world.

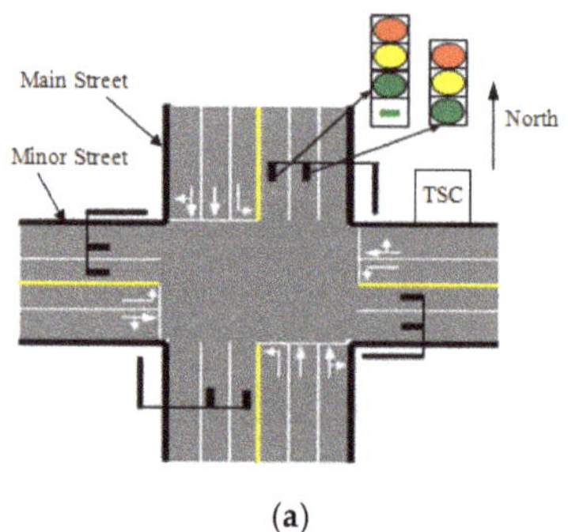

(a)

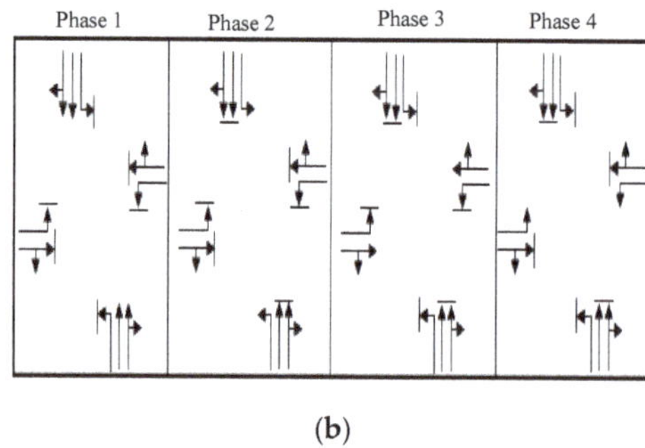

(b)

Figure 1. Traditional four-leg at-grade intersection model and its signal phase diagram: (**a**) Four-leg at-grade intersection model. (**b**) Intersection signal control phases. TSC: traffic signal controller.

This intersection model plays an important role in the safe passage of vehicles, but it cannot adapt to changes in traffic flow demand. For example, when there are fewer left-turning vehicles and more through vehicles, there will be congestion in the through lane, while the space of the exclusive left-turn lane will be wasted because through vehicles are forbidden in the exclusive left-turn lane. Irrespective of whether fixed-time control or an actuated traffic signal control are used at the intersection, it will cause the waste of left-turn phase duration and reduce the intersection's capacity, especially in the rush hour, which intensifies the traffic congestion and increases the overall delay of the vehicles. In practice, this is very common. For example, during the period of people going to work or returning home in areas near downtown, most vehicles are moving in and out of the city, and there are few or sometimes even no left-turning vehicles. In this case, the exclusive left-turn lane will reduce the capacity of the intersection and seriously affect the traffic's efficiency.

Therefore, this paper proposes a new intersection signal control model which changes the static lane direction arrow markings into variable lane direction arrow markings. The variable lane direction arrow markings can change with the number of vehicles, turning the exclusive left-turn lane into a controllable shared lane for left-turn and through movements. This can enable through vehicles to enter the controllable shared lane when there are fewer left-turning vehicles in the left-turn lane, thus solving the problem of wasting space in the exclusive left-turn lane. In addition, with the advantages of vehicle-to-infrastructure (V2I) communication, a control strategy is proposed for the new intersection signal control model which is based on a fixed-time control strategy and can solve the problem of full utilization of the green phase in the controllable shared lane. This provides the intersection with a better ability to adapt to changes in lane traffic flow, and thereby improves the intersection's capacity. The proposed intersection signal control model and control strategy were verified with the VISSIM simulator and analyzed in detail.

The rest of the paper is organized as follows. In Section 2, the research status of intersection signal control is discussed. Section 3 presents the new intersection signal control model and control principle. Section 4 proposes the new intersection signal control strategy. Numerical examples are presented and discussed for demonstration of the proposed intersection control model and algorithm in Section 5. Finally, Section 6 presents a brief conclusion and recommendations for future work.

2. Literature Review

In view of ameliorating the problem of the exclusive left-turn lane at intersections, several measures and treatments have been conceived, studied, and implemented. Hummer earlier summarized seven types of unconventional alternatives: median U-turn, bowtie, superstreet, paired intersections, jughandle, continuous flow intersection (CFI), and continuous green T; the pros and cons of each are described [10,11]. Other designs have also been proposed, such as the split intersection [12], USC intersection [13,14], paraflow intersection [15], etc. The aforementioned measures for the treatment of left turns are generally called unconventional arterial intersection designs (UAIDs), which are summarized and compared with each other in [16–19]. Reference to increasing intersection capacity and safety or reducing the delay at the intersection, as well as describing their application, have been discussed or validated in [20–24].

In addition, several new or improved methods have been developed. The drawbacks of CFI are summarized in [25], and the mid-block pre-signal method is given to overcome them, which is suitable for higher left-turn demand. The exit-lanes for left-turn (EFL) intersection has been presented, which opens up exit lanes for left-turning traffic dynamically with the help of an additional traffic light installed at the median opening, and it was found to effectively increase the capacity with a high level of application flexibility, especially under heavy left-turning traffic conditions [26]. Monte Carlo simulation was used to obtain optimal signal timings for a displaced left-turn (DLT) intersection, and it could provide near-optimum parameter selection ranges for the given traffic demands [27]. A generalized lane-based optimization model for the integrated design of DLT intersection types, lane markings, length of the displaced left-turn lane, and signal timings is presented, solved, simulated, and analyzed in [28]. Contraflow left-turn lanes (CLLs) are set up in the opposing lanes adjacent to the conventional left-turn lane at the intersection, and simulation analysis showed that CLLs outperformed a conventional left-turn lane design and generated less delay to both left-turn and through movements [29]. An unconventional U-turn treatment (UUT) for intersections which has a dual-bay design with different turning radii for small and large vehicles was presented, and could improve operations at intersection areas, especially when the volume/capacity ratio was small [30]. The traffic operational performance of three left-turn treatments under different traffic conditions was determined, and results showed that unconventional left-turn control types had less delay and travel time compared to the direct left-turn [31]. Moreover, a left-turn waiting area (LTWA) was used to reorganize left-turning traffic flows and to increase the capacity of signalized intersections, and simulation results showed that LTWA could improve the capacity for the left-turn movement [32,33].

Through the above-cited works, the intersection problem caused by left-turn lanes and green phase setting was improved. These works have increased the intersection capacity and/or safety or reduced the delay at the intersection, but they add more signals or need more land space. For example, the median U-turn removes left-turning movements from the major and minor approaches, forces left-turning drivers to proceed straight through the at-grade intersection and execute a U-turn at some distance downstream from the intersection location (directional crossover) in place of the traditional left-turn movement. During the procedure, signals are needed at the directional crossover and must be coordinated with the signal at the main intersection. Moreover, the improvements to the intersection are obtained only in heavy left-turning vehicles. Otherwise, the effects are not noticeable.

In recent years, the emerging connected vehicle (CV) and intelligent vehicle control technologies are likely to improve the safety, capacity, and operation efficiency of intersections. A CV can exchange data through real-time wireless communication including vehicle-to-vehicle (V2V), vehicle-to-infrastructure (V2I), and vehicle-to-device (V2D) using dedicated short-range communication (DSRC) protocols [34–36]. Intelligent vehicle control technology can realize intelligent driving [37,38], yaw control [39–41], and even automated driving [42,43].

Many related earlier works for traffic signal control have been accomplished by V2I or similar technology, and have achieved important results. An adaptive traffic light system based on wireless communication between vehicles and fixed controller nodes deployed in intersections was developed, and its improvement of traffic fluency and other clear advantages were validated via simulation [44]. A decentralized adaptive traffic signal control algorithm was given and simulated using supposed V2I communication data, and the different penetration rates of V2I vehicles were analyzed [45]. Also, an algorithm was proposed using information from connected vehicles to better adapt the traffic signal at an intersection with two one-way streets with on turns, and it was proven to be valuable [46]. A real-time adaptive signal phase allocation algorithm was presented utilizing vehicle location and speed data from connected vehicles, and optimized phase sequence and duration by solving a two-level optimization problem [47]. Using GPS trajectory data from a CV under low rates of market penetration, an approach to estimate traffic volume was developed and two case studies revealed that the approach could be of significant help to traffic management agencies for evaluating and operating traffic signals [48]. Under a partially connected and automated vehicles (CAVs) environment, an eco-driving system for an isolated signalized intersection was proposed to smooth out the shock wave caused by signal controls [49]. Given the assumption that advanced communication systems are available between vehicles and the traffic controller, arterial traffic signals for multiple travel modes were optimized and simulated [50]. A joint control framework for isolated intersections was investigated, modeled as a two-stage optimization problem, and validated as being able to reduce both vehicle delay and emissions compared to fixed-time and adaptive signal control [51]. A coordinated signal control system for urban ring roads under a vehicle–infrastructure connected environment was proposed and tested using a VISSIM simulation model to improve the average delay, number of stops, and queue length compared with a conventional traffic control system [52]. An optimal signal control algorithm using individual vehicle trajectory data under a V2I communication environment was developed and evaluated, showing superior performance to the actuated as well as fixed-signal control methods in an isolated intersection and a 2×3 signalized intersection network [53]. A new method for estimating the speed and position of non-connected vehicles at low CV penetration rates along a signalized intersection was developed and applied to the signal control strategy, and simulations in VISSIM showed the estimation accuracy to be higher for the intersection with fewer lanes [54]. An isolated intersection control problem is formulated as a game between the signal controller and the road users in the context of V2I, and numerical study showed that the method provided better performance and led to more even distribution of traffic than fixed-time control [55].

In addition, many studies have used V2I technology in the coordination control and optimization of intersections and vehicles to improve intersection operation efficiency. A method to process C2I communication data for tailback length approximation in urban networks was provided, and the tailback length could be used as a criterion to be optimized within signal control methods and used to provide an individual driver with optimal speed to pass the signalized intersection without stopping [56]. A cooperative method of traffic signal control and vehicle speed optimization for connected automated vehicles was proposed, and simulation showed a significant improvement of transportation efficiency and fuel economy [57]. Meanwhile, given the assumption of advanced communication technology between approaching vehicles and signal controller, a signal control algorithm was developed which allows for vehicle paths and signal control to be jointly optimized [58]. Three algorithms for traffic control using connected vehicles instead of stationary detectors were proposed (i.e., dynamic maximum gap, throughput-adjusted delay, and throughput-adjusted stopped time), and good performance was demonstrated [59]. Three categories of vehicles—conventional vehicles, connected vehicles, and automated vehicles—were considered, and an algorithm was proposed to find the optimal departure sequence to minimize the total delay based on position information. The simulation results indicated an evident decrease in the total number of stops and delay when using the connected vehicle algorithm for the tested scenarios, with information levels as low as 50% [60]. According to the controllability of CAVs, an innovative intersection operation scheme was proposed which can serve bi-directional traffic from one road in one signal phase, and which maximizes the intersection capacity by utilizing all lanes of the road at any given time [61].

The aforementioned studies applied information from V2I or similar technologies to improve the intersection signal control, and were all proved to be more efficient than conventional methods. However, there are few studies on V2I technology applied to left-turn signal control at intersections. Signalized left turn assist (SLTA) related to a cooperative intersection collision avoidance system (CICAS) was developed to provide information to left-turning drivers about the presence of oncoming vehicles based on proximity or available gap size [62]. SLTA does not apply more information to signal control at intersections. The information (i.e., vehicle location, type, and ID, etc.) based on V2I applied to improve the intersection problems caused by the exclusive left-turn lane space and the corresponding signal phase duration of the left-turn movements should be carefully discussed. In this paper, a new intersection control model is developed, and a control method using the information from V2I in a connected vehicle environment is proposed to resolve the problem caused by the exclusive left-turn lane space and corresponding signal phase duration, especially under the scenario of fewer left-turn movements during rush hours.

3. The New Intersection Signal Control Model and Control Principle

3.1. The New Intersection Signal Control Model

In order to solve the problem of lane space and phase duration waste caused by fewer or no left-turning vehicles in the left-turn lane, the exclusive left-turn lane is redesigned as a shared lane, in which the left-turning vehicles and through vehicles are permitted at the same time. The shared lane can be realized by changing the invariable lane direction arrow marking on the road surface to a variable one based on display technology (e.g., light-emitting diode, LED). Herein the intersection signal control model with the shared lane and variable lane direction arrow markings is named as the new intersection signal control model, in comparison to the traditional intersection (Figure 1a), and is shown in Figure 2.

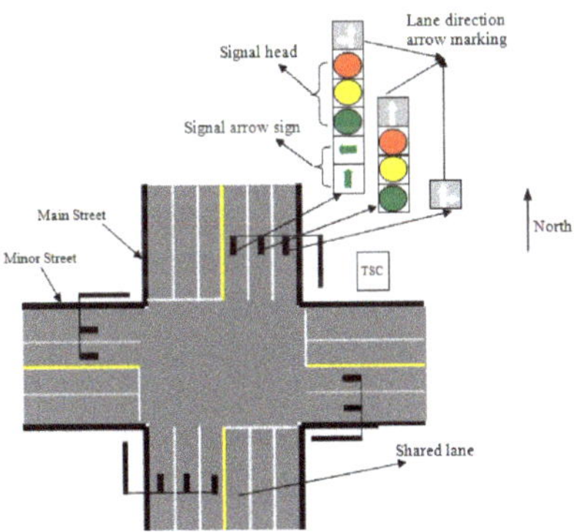

Figure 2. The new intersection signal control model diagram.

In Figure 2, instead of painting on the lane surface, lane direction arrow markings are mounted on a post above the corresponding lane with the signal heads and signal arrow signs. The lane direction arrow markings and signal arrow signs are designed using LED lights, which can display different signs. For example, the lane direction arrow marking above the shared lane can display left-turn, through, and through and left-turn arrow signs. When the lane direction arrow marking is a through and left-turn arrow sign, the left-turn and through vehicles can use this lane at the same time, which is just the function of a shared lane. In the new intersection signal control model, the shared through–right lane direction arrow markings can also display different signs, but these markings do not change in this paper, as the shared lane for through and left-turning vehicles and its markings are our research focus.

3.2. The Control Principle

In the new intersection signal control model, every vehicle has an onboard communication device, and the TSC (traffic signal controller) can communicate with them within the communication range of the V2I technology, by which the real-time vehicle movement information can be sent to the TSC. This information includes the vehicle type; the left-turn, right-turn, and through information; ID; position; and time stamp of each vehicle.

In addition, the TSC can analyze information from vehicles and decide the phase and phase duration of the next cycle based on the control strategy during the green signal time for the opposite direction. As in Figure 1, TSCs get and analyze information from vehicles during phase 3 and phase 4, and then decide the time duration of phase 1 and phase 2. In the scenario of less or no left-turn movements in the shared lane, all or some of the phase duration previously applied only for through movements can be used by the through vehicles in the shared lane. Furthermore, the part of the phase duration previously only for left-turn movements can also be used by through vehicles in the shared lane. Thus, the capacity of the intersection will be increased, and the performance of the intersection will be improved.

A diagram presenting the control principle of the new intersection signal control model and comparing it to the traditional intersection control model is shown in Figure 3.

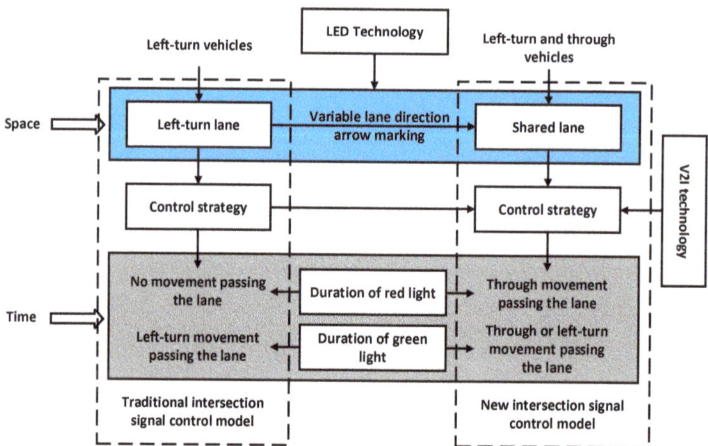

Figure 3. The control principle diagram. LED: light-emitting diode; V2I: vehicle-to-infrastructure.

4. The New Intersection Signal Control System and Control Strategy

4.1. Control System

The control system block diagram for the new intersection signal control model is shown in Figure 4. The control strategy is the key part and is discussed in Section 4.2.

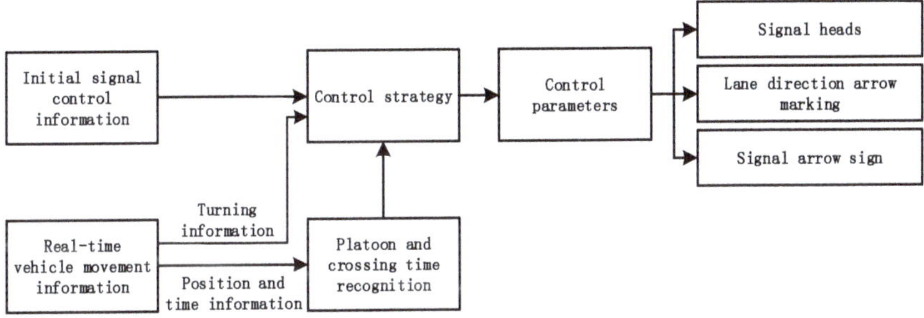

Figure 4. Block diagram of the control system.

Based on the control principle (Figure 3), the inputs of the control system include the fixed-time signal control parameters and the real-time vehicle movement information by V2I. The fixed-time signal control parameters, which are defined as the initial signal control information, include cycle length, phase and phase duration, yellow signal duration, and all-red signal duration.

The outputs of the control system are the control parameters for signal heads, lane direction arrow markings, and signal arrow signs. The control parameters for signal heads mainly include phase and phase duration. Here the cycle length, yellow signal, and all-red signal duration are supposed the same as initial signal control information. The control parameters for lane direction arrow markings include left-turn arrow, through arrow, or left-turn and through arrow. The control parameters for signal arrow signs denote that left-turn movements or through movements are permitted to pass the intersection.

4.2. Control Algorithm

4.2.1. The Initial Signal Control Information

The initial signal control information is the fixed-time control parameters based on [63,64], which were optimized and are suitable for the traffic flow dynamics of the day at the intersection (Figure 1), as shown in Figure 5. The variables are defined as follows:

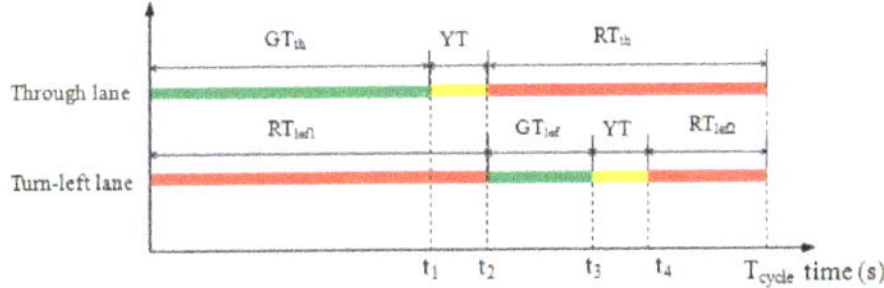

Figure 5. The northbound signal phases.

T_{cycle}: the signal cycle;
YT: yellow signal duration;
GT_{th}: through phase duration;
RT_{th}: red signal duration in the through lane;
GT_{lef}: left-turn phase duration;
RT_{lef1}, RT_{lef2}: red signal duration in the left-turn lane.

In addition, some time points to be used later in this paper are shown in Figure 5, such as t_1, t_2, t_3 and t_4, which can be expressed as Equation (1).

$$\begin{cases} t_1 = GT_{th} \\ t_2 = t_1 + YT \\ t_3 = t_2 + GT_{lef} \\ t_4 = t_3 + YT \end{cases} \tag{1}$$

4.2.2. Platoon Recognition and Its Passing Stop-Line Time Computation

In the new intersection signal control model, the through vehicle and left-turning vehicle coexist in the shared lane. If the time where they pass the stop line is obtained, the times RT_{lef1}, GT_{lef} can be used sufficiently. A method is proposed to calculate the passing stop-line time of vehicles in the shared lane based on the V2I information. The method includes platoon recognition and its passing stop-line time computation.

(1) Platoon Recognition

In the shared lane, the continuous and same-turn-signal vehicles are defined as a platoon. For example, in Figure 6, platoon I (PI) from the first to the nth vehicle are all through movement, but platoon II (PII) from the first to mth vehicles are all turning left. Platoon recognition is defined as the number of vehicles in the same platoon.

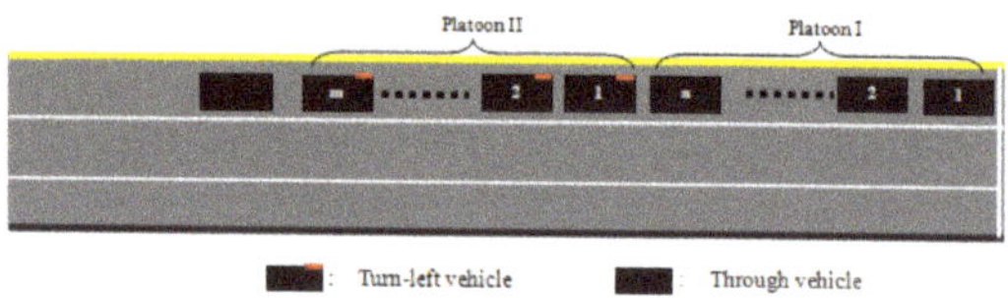

Figure 6. Platoons on the shared lane.

From Figure 5, during the time RT_{lef2}, vehicles will stop and queue up in the shared lane, and at the same time, the TSC will receive and analyze the real-time information from the vehicles in the shared lane. Based on the real-time information, the TSC obtains the left-turning and through vehicles' position distribution in the shared lane, recognizes the platoon, and obtains the number of vehicles and type of each vehicle in the platoon. For example, in Figure 6, PI includes the through vehicles from the first to the nth, and PII includes left-turning vehicles from the first to the mth, and the number of the PI and PII vehicles is n and m, respectively. Note that there may be different types of vehicles in the same platoon. In this situation, the number of vehicles in platoons should be converted to the number of standard cars according to the vehicle type. PI and PII in Figure 6 are assumed to be standard cars.

(2) Time of Platoon Passing the Stop Line

The time of the platoon passing the stop line at the intersection is related to many factors, such as driver behavior, reaction to the light, the intersection's topography, the headway of the platoon, etc. Some related research efforts have been made [65,66]. Because of the many factors related to the parameter, computation complexity is high. Besides, based on the V2I communication, a large amount of data about the time of vehicles passing the intersection is obtained easily, and one method is proposed here.

In the shared lane, once the green signal begins, the queued vehicles start and pass the intersection successively. Based on the vehicle position and time stamp, the TSC can obtain the time of vehicles passing the stop line, and then the number of vehicles passing the stop line can be computed according to the vehicle IDs. Therefore, the time of the platoon passing the stop line, in which the number of vehicles is recognized, can be calculated. Due to the differences between left-turning and through vehicles passing an intersection, the statistical data should be computed respectively.

Suppose the vehicles in the left-turn movement platoon are referred to as lef1, lef2, ..., lefn, the vehicles in the through movement platoon are th1, th2, ..., thm, and the corresponding time stamps of vehicles passing the stop line are $t_{lef1}, t_{lef2}, \ldots, t_{lefn}, t_{th1}, t_{th2}, \ldots, t_{thm}$, respectively.

Then, the time of the platoon passing the stop line can be obtained according to the start of green:

$$\begin{cases} T_{lef} = t_{lefn} - T_0, \\ T_{th} = t_{thm} - T_0, \end{cases} \tag{2}$$

where T_{lef} is the time of the left-turn movement platoon passing the stop line;

T_{th} is the time of the through movement platoon passing the stop line; and

T_0 is the start of green.

Thus, the time of all left-turn and through movement platoons passing the stop line can be obtained. Then, according to the obtained data, the average time τ_{lef}^N, τ_{th}^N of the platoon with the same number of left-turning and through vehicles passing the stop line can be calculated :

$$\begin{cases} \tau_{lef}^N = \text{AVERAGE}\left(T_{lef-N'}^1, T_{lef-N'}^2 \cdots\right), \\ \tau_{th}^N = \text{AVERAGE}\left(T_{th-N'}^1, T_{th-N'}^2 \cdots\right), \end{cases} \tag{3}$$

where N is the number of vehicles in the platoon, $N = 1,2,3, \dots$;

$T_{lef-N'}^1, T_{lef-N'}^2 \cdots$ are the time of the platoon with N left-turning vehicles to pass the stop line;

$T_{th-N'}^1, T_{th-N'}^2 \cdots$ are the time of the platoon with N through vehicles to pass the stop line.

According to this, the time series $\{\tau_{lef}^1, \tau_{lef}^2, \tau_{lef}^3, \dots\}$ and $\{\tau_{th}^1, \tau_{th}^2, \tau_{th}^3, \dots\}$ can be obtained. The time series is the average time of the left-turn movement platoon and through movement platoon passing the stop line under different numbers of vehicles in the shared lane at the intersection.

Then, according to the number of vehicles in the platoon recognition, the time of the platoon passing the intersection can be obtained by searching the time series. For example, if the number of vehicles in PI is N, the time T_{p1} can be obtained:

$$T_{p1} = \begin{cases} \tau_{lef}^N & \text{PI is the vehicles turning left} \\ \tau_{th}^N & \text{PI is the through vehicles} \end{cases} \tag{4}$$

Similarly, the time T_{P2} for PII can be obtained.

4.2.3. The Control Strategy

In the signal control system, once the control parameters are obtained, the TSC will drive the traffic light to fulfill the intersection control process. The control parameters are calculated by the control strategy, which is a key component of traffic signal control.

The inputs of the new intersection control system are initial signal control information and the real-time vehicle movement information. Based on Figure 5, the initial signal control information includes T_{cycle}, GT_{th}, YT, RT_{th}, RT_{lef1}, GT_{lef}, and RT_{lef2}.

On the basis of Formula (4), the real-time information includes the T_{p1} and T_{p2} for PI and PII, and their left-turn or through signal information, where *flag* = 0 means left-turning vehicle and *flag* = 1 means through vehicle. The real-time information can be expressed as $\text{PI}(T_{p1}, flag)$, $\text{PII}(T_{p2}, flag)$.

The output control parameters for signal heads above the through lane include green signal duration and red signal duration, labeled as NGT_{th}, NRT_{th}, respectively, and the parameters for the shared lane include green signal duration and red signal duration. Because left-turn and through vehicles coexist in the shared lane, the green and red signal durations may be composed of one or two segments, defined as NGT_{s1} and NGT_{s2} for the green signal, and NRT_{s1} and NRT_{s2} for the red signal. Among them, red signal duration $NRT_{s2} = RT_{lef2}$, in which the right-of-way is for opposite traffic flow, is constant.

From what has been discussed above, the control system can be expressed as the following mathematical expression:

$$[NGT_{th}, NRT_{th}, NGT_{s1}, NGT_{s2}, NRT_{s1}] = f \{T_{cycle}, GT_{th}, RT_{th}, RT_{lef1}, GT_{lef}, RT_{lef2}, YT, \\ \text{PI}(T_{p1}, flag), \text{PII}(T_{p2}, flag)\}. \tag{5}$$

The process for the control strategy is the following.

(1) *Flag* = 1 in $\text{PI}(T_{p1}, flag)$

This means that PI is composed of through vehicles, and the control strategy is as follows:

A: If $T_{p1} \geq t_3$, then

Shared lane: $NGT_{s1} = RT_{lef1} + GT_{lef}$, $NGT_{s2} = 0$, $NRT_{s1} = 0$, at green time NGT_{s1} the signal arrow sign is for through movement.

Through lane: $NGT_{th} = RT_{lef1} + GT_{lef}$, $NRT_{th} = RT_{lef2}$

Regarding the output signal control parameters, the control phases can be described as (a) in Figure 7.

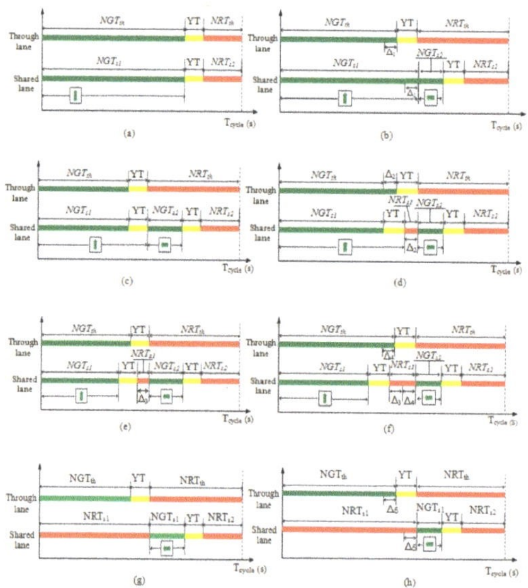

Figure 7. The output control phases under different conditions: (**a**) *Flag* = 1 and $T_{p1} \geq t_3$; (**b**) *Flag* = 1 in PI (T_{p1}, *flag*) and $t_2 \leq T_{p1} < t_3$; (**c**) *Flag* = 1 and $t_1 \leq T_{p1} < t_2$ and $T_{p2} \geq GT_{lef}$; (**d**) *Flag* = 1 and $t_1 \leq T_{p1} < t_2$ and $T_{p2} < GT_{lef}$; (**e**) *Flag* = 1 and $0 < T_{p1} < t_1$ and $T_{p2} \geq GT_{lef}$; (**f**) *Flag* = 1 and $0 < T_{p1} < t_1$ and $T_{p2} < GT_{lef}$; (**g**) *Flag* = 0 and $T_{p1} \geq GT_{lef}$; (**h**) *Flag* = 0 and $0 < T_{p1} < GT_{lef}$.

B: Else if $t_2 \leq T_{p1} < t_3$, then

Let $\Delta_1 = T_{p1} - t_2$.
Shared lane: $NGT_{s1} = RT_{lef1} + \Delta_1$, $NGT_{s2} = GT_{lef} - \Delta_1$, $NRT_{s1} = 0$, at the time of $NGTs_1$ the signal arrow signs are for through movements, and the time NGT_{s2} is for left-turn movements.

Through lane: $NGT_{th} = GT_{th} + \Delta_1$, $NRT_{th} = RT_{th} - \Delta_1$

Regarding the output signal control parameters, the control phase can be described as (b) in Figure 7.

C: Else if $t_1 \leq T_{p1} < t_2$, then

① If $T_{p2} \geq GT_{lef}$, then

Shared lane: $NGT_{s1} = GT_{th}$, $NGT_{s2} = GT_{lef}$, $NRT_{s1} = 0$, at the time of $NGTs_1$ the signal arrow signs is for through movements, and the time NGT_{s2} is for left-turn movements.

Through lane: the phase durations do not change, $NGT_{th} = GT_{th}$, $NRT_{th} = RT_{th}$.

Regarding the output signal control parameters, the control phase can be described as (c) in Figure 7.

② Else (means $T_{p2} < GT_{lef}$) then

Let $\Delta_2 = GT_{lef} - T_{p2}$.

Shared lane: $NGT_{s1} = GT_{th}$, $NGT_{s2} = GT_{lef} - \Delta_2$, $NRT_{s1} = \Delta_2$, at the time of NGT_{s1} the signal arrow sign is for through movements, and the time NGT_{s2} is for left-turn movements.

Through lane: $NGT_{th} = GT_{th} + \Delta_2$, $NRT_{th} = RT_{th} - \Delta_2$.

Regarding the output signal control parameters, the control phase can be described as (d) in Figure 7.

D: Else if $0 < T_{p1} < t_1$, then

Let $\Delta_3 = GT_{th} - T_{p1}$.

① If $T_{p2} \geq GT_{lef}$, then

Shared lane: $NGT_{s1} = GT_{th} - \Delta_3$, $NGT_{s2} = GT_{lef}$, $NRT_{s1} = \Delta_3$, at the time of NGT_{s1} the signal arrow sign is for through movements, and the time NGT_{s2} is for left-turn movements.

Through lane: the phases do not change, $NGT_{th} = GT_{th}$, $NRT_{th} = RT_{th}$.

Regarding the output signal control parameters, the control phase can be described as (e) in Figure 7.

② Else (means $T_{p2} < GT_{lef}$) then

Let $\Delta_4 = GT_{lef} - T_{p2}$.

Shared lane: $NGT_{s1} = GT_{th} - \Delta_3$, $NGT_{s2} = GT_{lef} - \Delta_4$, $NRT_{s1} = \Delta_3 + \Delta_4$. At the time of NGT_{s1} the signal arrow sign is for through movements, and the time NGT_{s2} is for left-turn movements.

Through lane: $NGT_{th} = GT_{th} + \Delta_4$, $NRT_{th} = RT_{th} - \Delta_4$.

Regarding the output signal control parameters, the control phase can be described as (f) in Figure 7.

(2) *Flag = 0* in the PI(T_{p1}, *flag*)

This means that PI is making left-turn movements, and the control strategy is as follows:

A: If $T_{p1} \geq GT_{lef}$ then

Shared lane: $NGT_{s1} = GT_{lef}$, $NGT_{s2} = 0$, $NRT_{s1} = RT_{lef1}$, at the time of NGT_{s1} the signal arrow sign is for left-turn movements.

Through lane: $NGT_{th} = GT_{th}$, $NRT_{th} = RT_{th}$.

Regarding the output signal control parameters, the control phase can be described as (g) in Figure 7.

B: Else (means $0 < T_{p1} < GT_{lef}$) then

Let $\Delta_5 = GT_{lef} - T_{p1}$.

Shared lane: $NGT_{s1} = GT_{lef} - \Delta_5$, $NGT_{s2} = 0$, $NRT_{s1} = RT_{lef1} + \Delta_5$, at the time of NGT_{s1} the signal arrow sign is for left-turn movements.

Through lane: $NGT_{th} = GT_{th} + \Delta_5$, $NRT_{th} = RT_{th} - \Delta_5$.

Regarding the output signal control parameters, the control phase can be described as (h) in Figure 7.

5. Simulation and Results

In this section, the new intersection signal control model and control strategy were evaluated based on an intersection located in Qingdao, China with a fixed-time signal control strategy. Relevant data and control parameters were obtained from the intersection. The cycle length was 129 s, the phases were the same as in Figure 1b, and the Phase 1 green signal duration was 48 s, Phase 2 was 25 s, Phase 3 was 25 s, and Phase 4 was 19 s. The intersection traffic state in rush hour was focused, in which very few vehicles were observed in the left-turn lane, a ratio of left-turning vehicles to total vehicles from 5% to 12.5% was observed, and the total vehicles were about 1800 per hour.

In this paper, VISSIM software was used as the simulation tool. The simulated intersection was designed and operated based on the data of the actual intersection. Firstly, the intersection model is shown in Figure 1a with an exclusive left-turn lane, and its control strategy was fixed-time. In the scenario, it is named as control algorithm I and the corresponding simulation process is called Simulation I.

Secondly, the lane direction arrow marking on the left-turn lane surface was painted as a through and left-turn arrow sign, as shown in Figure 8a. In this intersection model, the left-turning vehicles and through vehicles were in the same phase, so there were three phases of the intersection: Phase 1 for northbound and southbound left-turn and through movement, Phase 2 for the eastbound and westbound through movement, and Phase 3 for the eastbound and westbound left-turn movement, as shown in Figure 8b. Its control strategy was also fixed-time. The scenario is named as control algorithm II and the corresponding simulation process is called Simulation II.

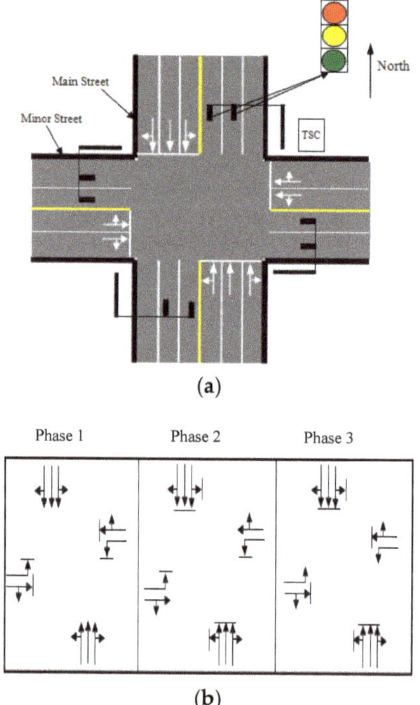

Figure 8. Control algorithm II intersection control model and its phases diagram: (**a**) Intersection control model. (**b**) Intersection signal control diagram.

Thirdly, the new intersection signal control model is shown in Figure 2 with shared lane, variable lane direction arrow markings. Its proposed control strategy is shown in Figure 7. The scenario is named as control algorithm III and the corresponding simulation process is called Simulation III.

The number of vehicles and the cycle length of the traffic control signal were the same in the three simulation processes. Because Simulation III had eight control results, the simulation time was set as $129 \times 8 = 1032$ s, and based on the total number of vehicles 1800 vehicle/h from the actual intersection, the numbers of left-turning vehicles of 100, 160, and 220 were selected.

In the simulation design, five signal groups in the signal controller for the intersection were considered, which had the same sequence (e.g., red–green–yellow). Signal group 1 controlled the through/right-turn vehicle movements for eastbound/westbound directions. Signal group 2 controlled the through movement for the shared lane. Signal group 3 controlled the through/right-turning vehicle movements for other directions. Signal group 4 controlled the left-turn movement for other directions. Signal group 5 controlled the turning movement for the shared lane.

In Simulation III, the shared lane was accessible to left-turning vehicles and through vehicles, the phase duration time was time-varying. To simulate the new signal control model and strategy (shown in Section 4.2.3), eight different periods were designed, which implemented different signal control strategies.

The results obtained included the number of vehicles passing the intersection (capacity of the intersection) with different input left-turning vehicles, and the percentage increase of Simulation III and II to Simulation I (Table 1). From Table 1, the capacity of the intersection increased in Simulations II and III, and Simulation III achieved the best results, meaning the proposed new signal control model and strategy could increase the performance of the intersection.

Table 1. The simulation results.

	Input Left-Turn Vehicles	Capacity of the Intersection	Increased Percentage
Simulation I	100	343	-
	160	345	-
	220	358	-
Simulation II	100	409	19.2%
	160	383	11%
	220	389	8.7%
Simulation III	100	440	28.3%
	160	442	28.1%
	220	430	20.1%

Besides, the three intersection signal control algorithms were simulated under different numbers of left-turning vehicles in the left-turn lane, and the capacities were obtained as shown in Figure 9.

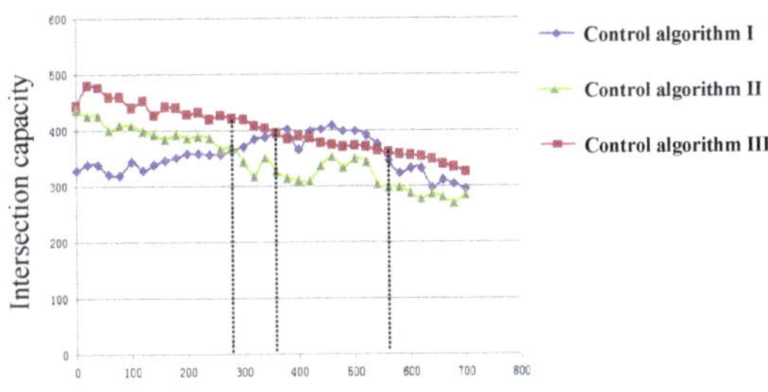

Figure 9. The capacities of intersection under three control algorithms.

From the results obtained in Figure 9, we can conclude the following:

(1) As the vehicles on the left-turn lane increased gradually, the intersection traffic capacity decreased under control algorithms II and III and increased for a long time under control algorithm I.

(2) When the number of vehicles in the left-turn lane was less than N_1, control algorithm III was the best for intersection traffic capacity, followed by control algorithm II, and the worst was control algorithm I. This means that control algorithms II and III are suitable for fewer left-turn movements. N_1 was the turning point of control algorithms I and II. Below N_1, control algorithm II was better, but after N_1 control algorithm I was better.

(3) N_2 and N_3 were the turning points of control algorithm I and control algorithm III. Below N_2, control algorithm III was better. Above N_2 and below N_3, control algorithm I was better than control algorithm III. After N_3, the control algorithm III was the best.

(4)   The control algorithm for the intersection can be given as follows:

① If the number of left-turning vehicles is less than N_2 or more than N_3, control algorithm III should be selected;

② If the number of left-turning vehicles is between N_2 and N_3, control algorithm I should be selected;

③ Under the condition of V2I technologies not being applied to the intersection control, control algorithm II should be used to improve the intersection traffic capacity, but its suitability term is the number of left-turning vehicles less than N_1.

Some discussions are as follows:

(1) In this paper, a new intersection signal control model is designed, the corresponding control strategy is proposed, and simulation showed that the new intersection signal control model and control strategy could improve the intersection capacity, especially in case of fewer left-turning vehicles in peak hours. However, other control models can be used in the case of more left-turning vehicles, as summarized in (4) above.

(2) The basic condition of the new intersection signal control model and strategy is that the left-turning and through vehicles line up in the shared lane as a platoon, but randomness is introduced when a vehicle without platoon control enters the shared lane. Due to fewer left-turning vehicles, the probability that left-turning and through vehicles will appear intermittently in a long period of time is impossible. Otherwise, it indicates that the number of left-turning and through vehicle is approximately equal, and the new intersection control model and algorithm is not suitable. However, in a short time period, it is possible that left-turning and through vehicles will appear intermittently, as shown in Figure 10.

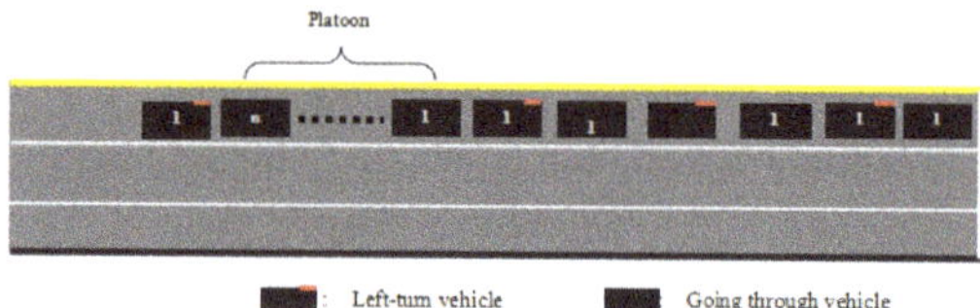

Figure 10. The through and left-turning vehicles appearing intermittently in the shared lane.

Under this case, it will be difficult for the control method proposed in this paper to improve the traffic efficiency of intersections. To solve the problem, a left-turn waiting area and a variable lane line were set up as shown in Figure 11.

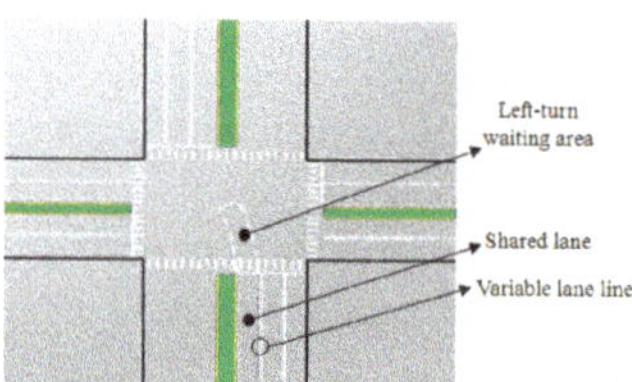

Figure 11. Diagram of left-turn waiting area and variable lane line for the case shown in Figure 10.

Left-turn waiting areas have been adopted in many urban intersections in China [32,33]. They are located at the front end of the left-turn lane (it is shared lane in this study) and extend into the interior of the intersection. The extension length should ensure that the left-turning vehicles waiting within this range do not conflict with the traffic flow in the opposite direction. Two parallel white dotted lines mark the left-turn waiting area, and the front end marks the stop line.

The design and working principle of the variable lane line are given in this paper. The variable lane line is composed of LED units, a transparent protective cover, a bearing box, the mask, and the control unit. The overall structure schematic diagram is shown in Figure 12.

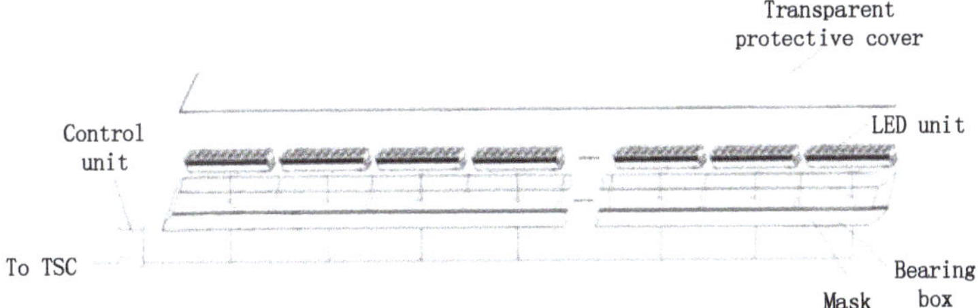

Figure 12. Schematic diagram of variable lane line structure.

The variable lane line LED unit groups are changeable, the number of groups in accordance with the length of the approach lane of the intersection, and connected to the TSC. The TSC can control the state of the variable line (e.g., on or off state), and its working diagram is shown in Figure 13. The variable lane line groups in Figure 13 consist of 13 LED units. Figure 13a demonstrates the working principle and effect of the variable lane line. When the LED is off, the variable lane color will be the same as the road surface (here it is shown in a different color to indicate its existence). Figure 13b represents the combination of variable lane line and conventional lane line with fixed marking, which is in a non-working state. In other words, this situation indicates that vehicles in the shared lane and its adjacent through lane cannot change lanes with each other. Figure 13c represents the combination of variable lane line and conventional lane line in a working state (i.e., through vehicles in the shared lane can change to the adjacent through lane if the through signal is allowed and lane changing is safe, but the vehicles in the through lane are prohibited from changing lanes to the shared lane).

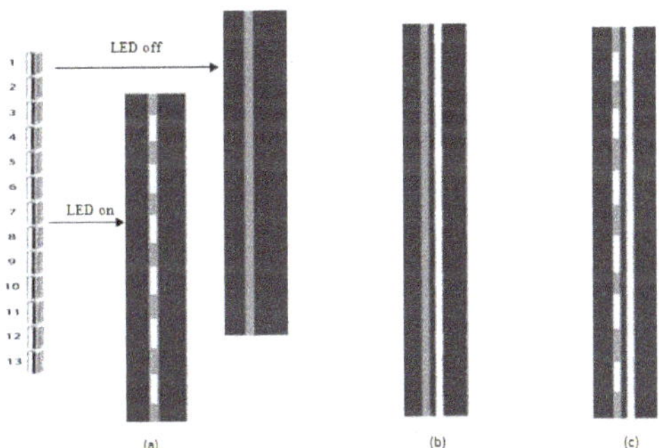

Figure 13. Schematic diagram of variable lane line: (**a**) Working principle and the effect of the variable lane line; (**b**) Non-working state of the combination of variable lane line and conventional lane line; (**c**) Working state of combination of variable lane line and conventional lane line.

In this circumstance, the control strategy of the shared lane will firstly give the through signal, and then the left-turn signal. As for the signal duration time for the through and left-turn movements, the storage capacity of the left-turn waiting area can be used as the basis for calculation. Assuming the time for the number of vehicles with the storage capacity to pass through the intersection is T_w, the left-turn phase duration time GT_{lef} must be greater than T_w (i.e., $GT_{lef} > T_w$), then the signal control phase diagram can be obtained under this condition, as shown in Figure 14. In the figure, $\Delta = GT_{lef} - T_w$, $NGT_{th} = GT_{th} + \Delta$, $NGT_{s1} = RT_{lef1} + \Delta$, $NGT_{s2} = T_w$. This control strategy is similar to the control phases (b) in Figure 7, and only the through and left-turn control durations are different (i.e., NGT_{th}, Δ, NGT_{s1}, and NGT_{s2} are different).

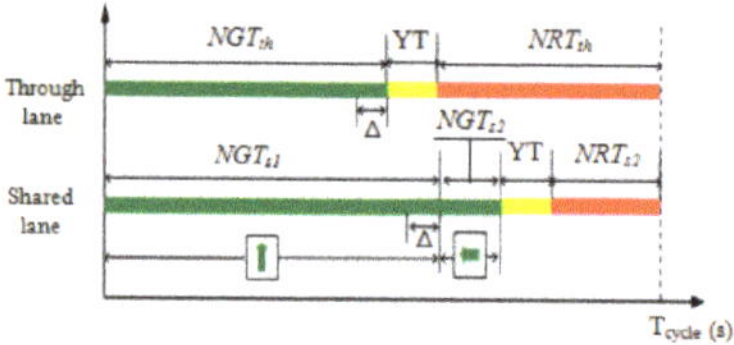

Figure 14. The control phases for the case shown in Figure 10.

Under this control strategy, the shared lane is firstly given the through signal, and then left-turn signal. At the through signal time the left-turning and through vehicles will move forward in turn. The through vehicles directly pass through the intersection and the left-turning vehicles will enter the left-turn waiting area. For example, for the vehicles in the shared lane of Figure 10, the first through vehicle will leave the stop line and pass the intersection directly, the second left-turning vehicle will enter the waiting area, then the third through vehicle will pass the intersection, and then the fourth left-turning vehicle will enter the waiting area. This continues in turn until the waiting area is full of left-turning vehicles. The process of the vehicle movements diagram is shown in Figure 15a.

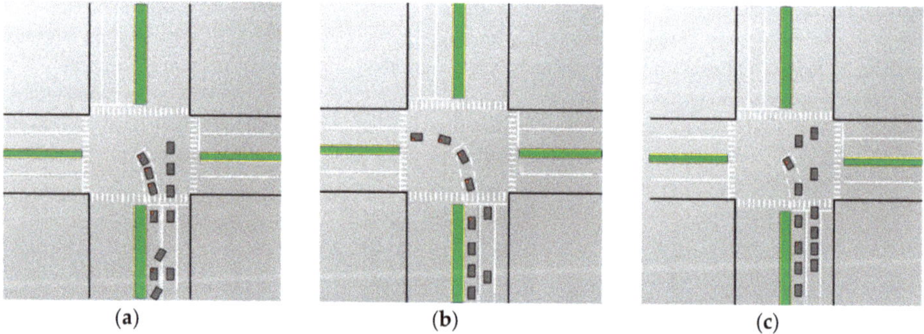

Figure 15. Vehicle movements diagram in shared lane and adjacent lane: (**a**) Unsaturated vehicle movements with the left-turn waiting area; (**b**) Filled vehicle movements with the left-turn waiting area; (c) Vehicle movements in left-turn green signal.

If there are no left-turning vehicles in the shared lane when the left-turn waiting area is full and the through green time is not ended, the through vehicles will pass through the intersection successively until the end of the through green time.

If there are still left-turning vehicles in the shared lanes, the left-turning vehicles will follow the through vehicles forward until stopping at the stop line, and this will inevitably hinder the subsequent through vehicles from passing the intersection. In this case, the variable lane line is activated (i.e., the lane line is in the working state; Figure 13c), which makes the through vehicles in the shared lane able to change lanes to the adjacent through lane, and then pass the intersection. The process of the vehicle movements diagram is shown in Figure 15b.

Moreover, when the green time for through movements is over, the green signal for left-turning movement is active, so the left-turning vehicles in the left-turn waiting area and the shared lane will pass through the intersection, while the through vehicles in the shared lane can move forward in turn to reach the stop line. The process of vehicles movements diagram is shown in Figure 15c.

6. Conclusions and Future Work

To improve the capacity of intersections with an exclusive left-turn lane in the case of fewer left-turning vehicles, especially in rush hour, a new intersection signal control model and control strategy are presented. This new intersection model includes a shared lane, realized by a variable lane direction arrow marking instead of markings painted on the surface of the road. Under the new intersection control model, a control strategy based on V2I communication technology is proposed.

To evaluate the intersection control model and strategy, an actual intersection was selected with field data collected, and micro-simulations of three different intersection control models and strategies with VISSIM software were conducted. The results show that the new intersection control model and strategy could increase the intersection capacity. The new intersection control model and strategy did not work well when the left-turning and through vehicles appeared intermittently in a short time period in the shared lane. This problem can be solved by setting up a variable lane line and left-turn waiting area at the intersection. In this circumstance, the control strategy was designed and the vehicle movements in different cases were discussed.

However, the new intersection signal control model and strategy was simulated with one-directional traffic, and the control strategy will be studied for all-direction traffic. Moreover, a strategy for the comprehensive application of variable lane direction arrow marking, variable lane lines, and shared lanes will be conducted to improve the intersection efficiency. Furthermore, some realistic constraints, such as the start time of the green phases, will be considered for further study.

Author Contributions: Methodology, C.R. and J.W.; Supervision, C.R. and Y.C.; Writing—original draft, J.W., L.Q. and S.L.; Writing—review and editing, C.R., and Y.C.

Funding: This research was funded by National Natural Science Foundation of China under grant number 71801144, Key Research and Department project of Shandong Province under grant number 2019GGX101008 and China Postdoctoral Science Foundation Funded Project under grant number 2019M652437.

Acknowledgments: Thanks to the editors and reviewers for their careful review, constructive suggestion and reminding, which helped improve the quality of the paper.

Conflicts of Interest: The authors declare no conflict of interest.

References

1. Desta, R.; Teklu, B.; Dananto, M. Simulation experiments and validation for considering bus priority treatments in small cites: The case of Hawassa city, Ethiopia. *Int. J. Eng. Sci. Technol.* **2018**, *10*, 21–28. [CrossRef]
2. Zhang, T.; Mao, B.; Chen, Z.; Ma, C.; Li, J. Research on bus signal priority strategy for urban arteries based on real-time vehicle queuing detection. In Proceedings of the IEEE 4th International Conference on Cloud Computing and Big Data Analysis, Chengdu, China, 12–15 April 2019.
3. Zhang, S.J.; Jia, S.P.; Bai, Y.; Mao, B.H.; Ma, C.R.; Zhang, T. Optimal adjustment schemes on the long through-Type bus lines considering the urban rail transit network. *Discret. Dyn. Nat. Soc.* **2018**. [CrossRef]
4. Hai, Y.; Tang, Y. Managing rail transit peak-Hour congestion with a fare-reward scheme. *Transp. Res. Part B* **2018**, *110*, 122–136.
5. Gao, G.; Sun, H.; Wu, J. Activity-Based trip chaining behavior analysis in the network under the parking fee scheme. *Transportation* **2019**, *46*, 647–669. [CrossRef]
6. Gao, G.; Sun, H.; Wu, J.; Liu, X.; Chen, W. Park-and-Ride service design under a price-Based tradable credits scheme in a linear monocentric city. *Transp. Policy* **2018**, *68*, 1–12. [CrossRef]
7. Sumalee, A.; Ho, H.W. Smarter and more connected: Future intelligent transportation system. *IATSS Res.* **2018**, *42*, 67–71. [CrossRef]
8. Cunha, F.; Maia, G.; Ramos, H.S.; Perreira, B.; Celes, C.; Campolina, A.; Rettore, P.; Guidoni, D.; Sumika, F.; Villas, L.; et al. Vehicular networks to intelligent transportation systems. *Emerg. Wirel. Commun. Netw. Technol.* **2018**, 297–315. [CrossRef]
9. Newell, G.F. Theory of highway traffic signals. *Res. Rep. Inst. Transp.* Available online: https://escholarship.org/uc/item/7zn2b9bc (accessed on 27 July 2017).

10. Hummer, J.E. Unconventional left-Turn alternatives for urban and suburban arterials—Part one. *ITE J.* **1998**, *68*, 26–29.

11. Hummer, J.E. Unconventional left-Turn alternative for urban and suburban arterials—Part two. *ITE J.* **1998**, *68*, 101–106.

12. Bared, J.G.; Kaisar, E.I. Benefits of split intersections. *Transp. Res. Rec. J.* **2000**, *1737*, 34–41. [CrossRef]

13. Tabernero, V.; Sayed, T.; Koscica, D. Introduction and analysis of new unconventional intersection scheme: Upstream signalized crossover intersection. In Proceedings of the 84th Annual Meeting of the Transportation Research Board, Washington, DC, USA, 9–13 January 2005.

14. Tabernero, V.; Sayed, T. Upstream signalized crossover intersection: An unconventional intersection scheme. *J. Transp. Eng.* **2006**, *132*, 907–911. [CrossRef]

15. Parsons, G.F. The parallel flow intersection: A new two-Phase signal alternative. *ITE J.* **2007**, *77*, 28–37.

16. Rodegerdts, L.A.; Nevers, B.L.; Robinson, B. *Signalized intersections: Informational Guide*; FHWA-HRT-04-091; Federal Highway Administration: Washington, DC, USA, 2004.

17. Hughes, W.; Jagannathan, R.; Sengupta, D.; Hummer, J. *Alternative Intersections/Interchanges: Informational Report (AIIR)*; FHWA-HRT-09-060; Federal Highway Administration: Washington, DC, USA, 2010.

18. El Esawey, M.; Sayed, T. Analysis of unconventional arterial intersection designs (UAIDs): State-of-the-Art methodologies and future research directions. *Transportmetrica* **2013**, *9*, 860–895. [CrossRef]

19. ATTAP, University of Maryland, Unconventional Intersection Design & Strategies. Available online: http://attap.umd.edu/uaid.php (accessed on 18 September 2018).

20. Bared, J.G.; Kaisar, E.I. Median U-Turn design as an alternative treatment for left turns at signalized intersections. *ITE J.* **2002**, *72*, 50–54.

21. Leng, J.; Zhang, Y.; Sun, M. VISSIM-Based simulation approach to evaluation of design and operational performance of U-Turn at intersection in China. In Proceedings of the 2008 International Workshop on Modelling, Simulation and Optimization, Hong Kong, China, 27–28 December 2008.

22. Liu, P.; Qu, X.; Wang, W.; Cao, B. Using VISSIM to model capacity of U-Turns at unsignalized intersections with non-traversable median cross sections. In Proceedings of the Transportation Research Board 89th Annual Meeting, Washington DC, USA, 10–14 January 2010.

23. Dhatrak, A.; Edara, P.; Bared, J.G. Performance analysis of parallel flow intersection and displaced left-Turn intersection designs. *Transp. Res. Rec.* **2010**, *2171*, 33–43. [CrossRef]

24. Elazzony, T.; Talaat, H.; Mosa, A. Microsimulation approach to evaluate the use of restricted lefts/through U-Turns at major intersections—A Case study of cairo-Egypt urban corridor. In Proceedings of the 13th International IEEE Annual Conference on Intelligent Transportation Systems, Madeira Island, Portugal, 19–22 September 2010.

25. Xuan, Y.; Daganzo, C.F.; Cassidy, M.J. Increasing the capacity of signalized intersections with separate left turn phases. *Transp. Res. Part B* **2011**, *45*, 769–781. [CrossRef]

26. Zhao, J.; Ma, W.; Zhang, H.; Yang, X. Increasing the capacity of signalized intersections with dynamic use of exit lanes for left-Turn traffic. *Transp. Res. Rec.* **2013**, *2355*, 49–59. [CrossRef]

27. Suh, W.; Hunter, M.P. Signal design for displaced left-turn intersection using Monte Carlo method. *KSCE J. Civ. Eng.* **2014**, *18*, 1140–1149. [CrossRef]

28. Zhao, J.; Ma, W.; Head, K.L.; Yang, X. Optimal operation of displaced left-Turn intersections: A lane-Based approach. *Transp. Res. Part C* **2015**, *61*, 29–48. [CrossRef]

29. Wu, J.; Liu, P.; Tian, Z.Z.; Xu, C. Operational analysis of the contraflow left-Turn lane design at signalized intersections in China. *Transp. Res. Part C* **2016**, *69*, 228–241. [CrossRef]

30. Xiang, Y.; Li, Z.; Wang, W.; Chen, J.; Wang, H.; Li, Y. Evaluating the operational features of an unconventional dual-Bay U-Turn design for intersections. *PLoS ONE* **2016**, *11*, e0158914. [CrossRef] [PubMed]

31. Taha, M.M.A.; Abdelfatah, A. Impact of using indirect left-Turns on signalized intersections' performance. *Can. J. Civ. Eng.* **2015**, *44*, 462–471. [CrossRef]

32. Yang, Z.; Liu, P.; Chen, Y.; Yu, H. Can left-Turn waiting areas improve the capacity of left-Turn lanes at signalized intersections. *Procedia-Soc. Behav. Sci.* **2012**, *43*, 192–200. [CrossRef]

33. Ma, W.; Liu, Y.; Zhao, J.; Wu, N. Increasing the capacity of signalized intersections with left-Turn waiting areas. *Transp. Res. Part A* **2017**, *105*, 181–196. [CrossRef]

34. Research and Innovative Technology Administration. Available online: http://www.its.dot.gov/connected_vehicle.htm (accessed on 6 September 2013).

35. USDOT. Connected Vehicle Reference Implementation Architecture. Available online: http://www.iteris.com/cvria/html/applications/applications.html (accessed on 27 July 2015).
36. USDOT. ITS Research Fact Sheets. Available online: http://www.its.dot.gov/communications/its_factsheets.htm (accessed on 27 June 2016).
37. Jin, N. Design of visual feature detection system for intelligent driving of electric vehicle. In Proceedings of the IEEE International Conference on Robots & Intelligent System, Changsha, China, 26–27 May 2018.
38. Yu, L.; Shao, X.; Wei, Y.; Zhou, K. Intelligent land-Vehicle model transfer trajectory planning method based on deep reinforcement learning. *Sensors* **2018**, *18*, 2905.
39. El Hajjaji, A.; Chadli, M.; Oudghiri, M.; Pagès, O. Observer-Based robust fuzzy control for vehicle lateral dynamics. In Proceedings of the American Control Conference, Minneapolis, MN, USA, 14–16 July 2006.
40. Mi, T.; Li, C.; Hu, C.; Wang, J.; Chen, N.; Wang, R. Robust H∞ output-Feedback yaw control for in-Wheel motor driven electric vehicles with differential steering. *Neurocomputing* **2016**, *173*, 676–684.
41. Dahmani, H.; Chadli, M.; Rabhi, A.; El Hajjaji, A. Vehicle dynamics and road curvature estimation for lane departure warning system using robust fuzzy observers: Experimental validation. *Veh. Syst. Dyn.* **2015**, *53*, 1135–1149. [CrossRef]
42. Milz, S.; Arbeiter, G.; Witt, C.; Abdallah, B.; Yogamani, S. Visual SLAM for automated driving: Exploring the applications of deep learning. In Proceedings of the IEEE Conference on Computer Vision and Pattern Recognition Workshops, Salt Lake City, UT, USA, 18–22 June 2018. [CrossRef]
43. Hubmann, C.; Schulz, J.; Becker, M.; Althoff, D.; Stiller, C. Automated driving in uncertain environments: Planning with interaction and uncertain maneuver prediction. *IEEE Trans. Intell. Veh.* **2018**, *3*, 5–17. [CrossRef]
44. Gradinescu, V.; Gorgorin, C.; Diaconescu, R.; Cristea, V.; Iftode, L. Adaptive traffic lights using car-to-Car communication. In Proceedings of the IEEE 65th Vehicular Technology Conference, Dublin, Ireland, 22–25 April 2007.
45. Priemer, C.; Friedrich, B. A decentralized adaptive traffic signal control using V2I communication data. In Proceedings of the 12th International IEEE Conference on Intelligent Transportation Systems, St. Louis, MO, USA, 4–7 October 2009.
46. Guler, S.I.; Menendez, M.; Meier, L. Using connected vehicle technology to improve the efficiency of intersections. *Transp. Res. Part C* **2014**, *46*, 121–131. [CrossRef]
47. Feng, Y.; Head, K.L.; Khoshmagham, S.; Zamanipour, M. A real-Time adaptive signal control in a connected vehicle environment. *Transp. Res. Part C* **2015**, *55*, 460–473. [CrossRef]
48. Zheng, J.; Liu, H.X. Estimating traffic volumes for signalized intersections using connected vehicle data. *Transp. Res. Part C* **2017**, *79*, 347–362. [CrossRef]
49. Jiang, H.; Hu, J.; An, S.; Wang, M.; Park, B.B. Eco approaching at an isolated signalized intersection under partially connected and automated vehicles environment. *Transp. Res. Part C* **2017**, *79*, 290–307. [CrossRef]
50. He, Q.; Head, K.L.; Ding, J. PAMSCOD: Platoon-Based arterial multi-Modal signal control with online data. *Transp. Res. Part C* **2012**, *20*, 164–184. [CrossRef]
51. Feng, Y.; Yu, C.; Liu, H.X. Spatiotemporal intersection control in a connected and automated vehicle environment. *Transp. Res. Part C* **2018**, *89*, 364–383. [CrossRef]
52. Ma, C.; Hao, W.; Wang, A.; Zhao, H. Developing a coordinated signal control system for urban ring road under the vehicle-Infrastructure connected environment. *IEEE Access* **2018**, *6*, 52471–52478. [CrossRef]
53. Han, E.; Pil Lee, H.; Park, S.; So, J.; Yun, I. Optimal signal control algorithm for signalized intersections under a V2I communication environment. *J. Adv. Transp.* **2019**. [CrossRef]
54. Kamal, K.; Chandan, K. A real-Time traffic signal control strategy under partially connected vehicle environment. *Promet Traffic Transp.* **2019**, *31*, 61–73.
55. Xu, Y.; Li, D.; Xi, Y.; Lin, S. Game-Based traffic signal control with adaptive routing via V2I. In Proceedings of the IEEE International Conference on Intelligent Transportation Systems, Maui, HI, USA, 4–7 November 2018.
56. Christian, P.; Bernhard, F. A method for tailback approximation via C2I-Data based on partial penetration. In Proceedings of the 15th World Congress on Intelligent Transport Systems and ITS America's 2008 Annual Meeting, New York, NY, USA, 16–20 November 2008.

57. Xu, B.; Jeff, B.X.; Yougang, B.; Wan, L.; Wang, J.; Eben, LS.; Li, K. Cooperative method of traffic signal optimization and speed control of connected vehicles at isolated intersections. *IEEE Trans. Intell. Transp. Syst.* **2019**, *20*, 1390–1403. [CrossRef]

58. Li, Z.; Elefteriadou, L.; Ranka, S. Signal control optimization for automated vehicles at isolated signalized intersections. *Transp. Res. Part C* **2014**, *49*, 1–18. [CrossRef]

59. Pereira, A.M. Traffic signal control for connected and non-Connected vehicles. In Proceedings of the Smart City Symposium Prague, Prague, Czech Republic, 24–25 May 2018.

60. Yang, K.; Guler, S.I.; Menendez, M. Isolated intersection control for various levels of vehicle technology: Conventional, connected, and automated vehicles. *Transp. Res. Part C* **2016**, *72*, 109–129. [CrossRef]

61. Sun, W.; Zheng, J.; Liu, H.X. A capacity maximization scheme for intersection management with automated vehicles. *Transp. Res. Procedia* **2018**, *23*, 121–136. [CrossRef]

62. Page, T.R.D. *Multiple Sources of Safety Information from V2V and V2I: Redundancy, Decision Making, and Trust-Safety Message Design Report*; U.S. Department of Transportation: Washington, DC, USA, 2015.

63. Webster, F.V. Traffic signal settings. In *Road Research Technique Paper*; Road Research Laboratory: London, UK, 1958.

64. Khisty, C.J.; Lall, B.K. *Transportation Engineering: An Introduction*, 3rd ed.; Pearson: London, UK, 2002.

65. Gartner, N.H. Development and implementation of an adaptive control strategy in a traffic signal network: The virtual-fixed-cycle approach. In *Transportation and Traffic Theory in the 21st Century, Proceedings of the 15th International Symposium on Transportation and Traffic Theory, Adelaide, Australia, 16–18 July 2002*; Emerald Group Publishing Limited: Bingley, UK, 2002.

66. Akcelik, R.; Besley, M.; Roper, R. *Fundamental Relationships for Traffic Flows at Signalized Intersections*; ARRB Transportation Research Ltd.: Vermont South, VIC, Australia, 1999.

MDPI
St. Alban-Anlage 66
4052 Basel
Switzerland
Tel. +41 61 683 77 34
Fax +41 61 302 89 18
www.mdpi.com

Electronics Editorial Office
E-mail: electronics@mdpi.com
www.mdpi.com/journal/electronics

www.ingramcontent.com/pod-product-compliance
Lightning Source LLC
LaVergne TN
LVHW070703170726
843515LV00004B/468